BRITISH AND AMERICAN TANKS OF WORLD WAR II

1. Most important Allied tank of the Second World War was the M4 Medium (Sherman). This M4A1 of the US Marines comes ashore in typical scenery in the Pacific campaign, 1943.

BRITISH AND AMERICAN TANKS OF WORLD WAR II

The complete illustrated history of British, American and Commonwealth tanks, gun motor carriages and special purpose vehicles, 1939–1945

by
Peter Chamberlain and Chris Ellis

ARMS AND ARMOUR PRESS

Published by
ARMS AND ARMOUR PRESS
Lionel Leventhal Limited
677 Finchley Road
Childs Hill
London, N.W.2

First published 1969

SBN 85368 033 7

Printed and Bound in Great Britain by
The Garden City Press Limited,
Letchworth, Hertfordshire

CONTENTS

2. First of a new breed of purpose-built American tank destroyers was the M10 Gun Motor Carriage.

PREFACE

DESPITE the growing number of publications devoted to the tanks and armoured vehicles of World War II, no single volume has hitherto attempted to describe *all* the tanks and their armoured variants used by the British, American, and main British Commonwealth armies in the years 1939-45. This book, *British and American Tanks of World War II,* does just that, however, and offers the most comprehensive coverage of the subject yet undertaken. It sets on record the hundreds of different tank types and variants developed in that tumultuous period of history when tanks almost literally "came of age"; when in less than five years what was general grew from tiny lightly armed vehicles commonly little bigger than the average motor car to complex forty ton monsters, powerfully armed and heavily armoured, all under the influence of the strongest impetus to change known to man—war.

In 1939 both Britain and the United States had inadequate tank forces. In both cases tank doctrine had progressed little since 1918 and equipment was still geared to a mixture of vehicles for infantry support (and trench warfare), and vehicles for the "mechanised cavalry" role, neither type being really compatible. When the crunch came, swiftly and dramatically, with the German invasion of the Low Countries and France in May 1940, both Britain and the United States were faced with the need for tanks quickly and in prodigious numbers. This led to a massive increase in the industrial participation in tank production, and the appearance of dozens of new designs in quick succession from then on to meet changing tactical requirements and the constant need to produce the right vehicle for the job. Both Britain and the United States also had to develop special purpose vehicles (like bridgelayers, mine clearers and so on), so that in some cases there might eventually be more than a hundred versions, counting all the improved production types, of one basic vehicle.

The scope of this book

To keep this book as simple as possible we have first of all grouped all vehicles by nation of origin—British first, then American, then Commonwealth nations. Within this broad framework we have further subdivided vehicles into classes in order of size, giving, for example "British Light Tanks", "British Cruiser Tanks", and so on. Within each section so formed each type is presented in chronological order—earliest first. All further variants or special purpose versions developed on any given chassis are described immediately following the basic vehicle, whether or not the variants concerned were re-classified for another role quite different from that of the original vehicle. Self-propelled motor carriages which were standardised and saw full service are given a separate entry—still following the entry of the basic vehicle, however. The Infantry Tank Mk III, Valentine, provides a good example of this. It comes first under the section heading "British-built Infantry Tanks", where it follows the Infantry Tanks Mks I and II, and starts with an entry devoted to the basic Valentine and the various models of it developed as an infantry tank. These are followed by the special purpose variants (like mine clearers and "swimming" tanks), and finally by experimental variants of the chassis. Two self-propelled gun variants which each had a development history of their own, the Bishop and Archer, are given separate entries, however, immediately following the coverage of the Valentine proper. In the American sections vehicles which saw British service each carry an additional section devoted to this and any variants developed specifically by the British, and not of

American origin, are described under the "British Service" heading. Thus the British Sherman Firefly is found in the British Service section under the M4 series heading, which itself comes under the main section devoted to American-built medium tanks. Thus essentially this book describes development by chassis and not by function. However, the index (Appendix 5) is divided by function, so that all British self-propelled guns, for example, are grouped together under that heading with appropriate cross-reference to the pages concerned. There are one or two exceptions to the foregoing, mainly in the cases of vehicles with only very few special purpose derivatives, where even "standardised" SP types can be described under the entry devoted to the basic tank. Thus in the M24 Chaffee entry, the "Light Weight Combat Team" of motor carriages can be very conveniently grouped immediately after the M24 itself, mainly to save space.

Specifications and data

Each entry contains a concise history of the vehicle concerned, with facts relevant to its development, and as all the vehicles are grouped as nearly as possible in chronological order, the book from beginning to end gives a fairly continuous development narrative for the tanks of each nation covered. We considered this a more interesting approach than the alternative one of several chapters of generalised development history remote from the actual vehicle entries themselves. Brief specifications are given for each vehicle, which include all salient technical details. To save space, however, armament and engine details are kept to the minimum in the specifications but are dealt with more fully in Appendix I. There are numerous cross-references in the text—indicated as (qv)—which are made necessary by the complexity of the overall development history of British and American tanks and by the fact that numerous vehicles were being developed concurrently. Any cross-reference marked is mainly intended as extra guidance for the reader, suggesting that he turns back to the entry indicated for more details on what might be only a passing mention on the page in hand.

All weights in the specifications are given in pounds only, due to the discrepancy between the British and American tons, although in the text weights are occasionally quoted in tons but qualified as either (long) (British) or (short) (American). Armour thicknesses are quoted in millimetres throughout and only maximum and minimum are given. An inch is taken to equal 25 millimetres. Even though there are more than 500 illustrations in this book, it has not been possible to show all variants described. To utilise space to the best advantage we have avoided duplicating pictures of vehicles which looked essentially the same externally. Thus, because the Cavalier, Centaur, and Cromwell ARVs all looked alike we have shown only one and cross-referenced the other two to the one picture. A few very minor experimental variants have been omitted completely.

Illustrations

In selecting the illustrations for this book we have, as far as possible, used views of each type from several different aspects to give the reader the fullest idea of all-round details and appearance. Though we have used the best available photographs, in some cases illustrations of the rarer and more obscure vehicles are of less than perfect quality, but these are frequently the only remaining pictorial record of a vehicle's appearance; enthusiasts will, we know, understand this. Appendix 5 comprises a full index, and this gives plate numbers and the page numbers on which descriptions of all the 569 illustrations in this book will be found. For this reason there is not, of course, a separate list of illustrations by number.

Acknowledgments

In compiling this book we have been aided by several individuals who are experts in their various fields of interest and who have either provided pictures or extra information, or both, to supplement our own knowledge. Particular thanks go to Col R. J. Icks, USAR (Retd), Major J. Loop, US Army, and Richard J. Hunnicutt, for assistance with coverage of the American vehicles. Mr Hunnicutt also provided much useful data, plus most of the American production figures. Col Icks provided many of the pictures of the more obscure American vehicles. John Milsom gave invaluable assistance with details of the more obscure British variants, and further information on some British vehicles came from Raymond Surlémont. Thanks are also due to Messrs J. Golding, G. Pavy, and E. Hine of the Imperial War Museum photographic library for assistance with pictures, and to Miss R. Coombs, librarian, and Mr D. Nash of the Imperial War Museum reference library for assistance in locating archive material. Finally our thanks to Major-Gen N. W. Duncan and Col P. H. Hordern, past and present Curators respectively of the Royal Armoured Corps Tank Museum, Bovington, Dorset, for research facilities. Many vehicles described in this book can be seen in the Royal Armoured Corps Tank Museum, which is open to the public daily. Lastly our thanks go to Kenneth M. Jones, who supplied the excellent drawings given in Appendix 2. Picture sources for this book are as follows: Imperial War Museum, Chamberlain Collection, Icks Collection, Hunnicutt Collection, Aberdeen Proving Ground, US Signal Corps, Canadian Official, Australian Official.

Peter Chamberlain
Chris Ellis

3. The best British tank in service in 1940 was the Infantry Tank Mk II, Matilda, which was almost immune to German and Italian anti-tank and tank guns then in use. The Matilda spearheaded British armour in the successful Libyan campaign of late 1940, when these vehicles were pictured.

INTRODUCTION

IN this book the major factors which directly influenced the design and evolution of any specific vehicle are given in detail in the development histories of the individual vehicles. It is desirable, however, to outline here background summary of development authorities, trends and changes which characterised British and American tank development in the 1939-45 period. It must be emphasised that although the United States were not directly involved in armed conflict until December 1941, American weapons development, including that of tanks, was greatly influenced by observation of the fortunes of war in the years 1939-40, while orders for tanks to supplement depleted British stocks were also placed in 1940 on a "cash and carry" basis, long before the Lend-Lease system came into operation and made further vehicles available.

Development and Procurement authorities

In Britain prior to 1936 the Master General of the Ordnance was the supreme authority responsible for tank design and procurement. Under him the Director of Mechanisation supervised actual design work in conjunction with the Mechanisation Board, which was a committee made up of senior representatives of the "user" arms. By the outbreak of war in 1939, the Master General of the Ordnance had become the Director General of Munitions Production and all designs and procurement responsibilities were transferred from the War Office to the newly-established Ministry of Supply. Overall tank design responsibility then came under the Director-General of Tanks and Transport with, in 1940, a Controller of Mechanisation supervising the Director of Mechanisation, who worked with the Mechanisation Board as before. In May 1940, following British reverses in the French campaign, a new War Cabinet was formed under Winston Churchill, who approved the setting up of a Tank Board to examine faults in the existing design and procurement system and to advise on improvements. They proposed a Director of Armoured Fighting Vehicles (DAFV) to represent the War Office (General Staff) interest, with separate Directors of Design and Production, all under the Director-General of Tanks and Transport, who took the place of the old Director of Mechanisation.

Early in 1941 the Tank Board was reorganised and given executive powers to expedite War Office requirements in matters affecting tanks. Included on the board were the Director-General of Tanks and Transport and the Director of Artillery (for tank gun, anti-tank gun, ammunition, and SP equipment matters), plus DAFV and General Staff representatives. In September 1942 a Chairman, Armoured Fighting Vehicles Division, was appointed, who also became chairman of the Tank Board and was the chief executive responsible for tank design in the Ministry of Supply. The Tank Board was also reconstituted to contain equal representation from the Ministry of Supply and the War Office (who represented the "users"). This general organisation remained in force until the end of the war.

British design authorities

On the design side itself, however, there were several important changes largely due to the vast industrial participation in tank production, which had increased dramatically since the outbreak of the war. Such vehicles as the Churchill and Valentine, for example, were designed mainly by the firms which built them, with only relatively minor help from the Department of Tank Design, the organisation, which, following the 1940 reforms, carried out actual design work under the Director-General of Tanks and Transport. In late 1941 the Department of Tank Design was placed under the Controller General of Research and Development, and as the war progressed the department changed its function from designing proper to co-ordination of design and production facilities. In other words, instead of actually designing a vehicle itself, the Department of Tank Design passed requirements to one of the tank producers and approved (and if necessary improved) the design the producers drew up. The old "drawing board" orders, which had generally resulted in tanks (like the Churchill and Covenanter) with a formidable record of "teething troubles", became a thing of the past. Under the new organisation at least six pilot models were generally built. Similarly, a proper "design parentage" organisation was built up whereby one particular company took full charge of design and production of one particular vehicle and supervised all necessary subcontract work for the vehicle in question. The Churchill (Vauxhall) and Valentine (Vickers) in 1940 set this pattern, subsequently adopted with all later British tanks, and the Department of Tank Design did not itself design a complete tank again until 1944-45, when it was responsible for the Centurion. By 1945 the Department had become very influential indeed and, in the circumstances, left a most creditable wartime record in the face of continually fluctuating War Office (ie "user") requirements, frequent friction between War Office and Ministry of Supply, and a good deal of War Office conservatism.

American design authorities

In the United States the division of responsibility between the "users" and the design and development side was more clearly defined. After World War I, tanks became an infantry responsibility in the US Army, but in the thirties the US Cavalry was also equipped with tanks in the guise of "combat cars" (for details of circumstances, see Combat Car M1 entry). In July 1940 tank units were taken away from these two "user" arms and organised into the Armored Force, while a Tank Destroyer Command was formed to operate self-propelled guns. In March 1942, the US War Department was reorganised and all "using arms", including the Armored Force and Tank Destroyer Command, became part of Army Ground Forces (AGF). At the same time the various departments which dealt with supply and procurement throughout the army were merged to become the Services of Supply, later called the Army Service Forces (ASF). Virtually unchanged by the reorganisation was the Ordnance Department, although it lost its power to influence the General Staff directly on procurement matters, and became instead dependent on the ASF for procurement authority. The Ordnance Department was responsible for all design and development of US Army ordnance items and, as its name implies, dated from the days when artillery was the main weapon. Tanks, of course, were included among the Ordnance Department's design responsibilities. There was a two-way transfer of ideas between the "users", AGF, and the Ordnance Department, in that Ordnance could suggest new equipment and propose new designs to AGF, while AGF could ask Ordnance to design or produce ideas to meet their requirements. There was a secondary line of communication between the Ordnance Department and overseas theatres of war on a similar basis. Within AGF, the Armored Force Board looked after tank matters. Approval for fundings for procurement of new equipment meanwhile had to come from ASF. It is essential to appreciate these relationships (presented here in much simplified form) in order to understand many of the references to them in development histories of the American vehicles described later.

In theory this organisation was perfectly simple, but as in Britain there was frequent friction between the "users" and the "suppliers", generally involving the forces of change in the form of the designers in the Ordnance Department against the forces of conservatism in the form of the Generals in Army Ground Forces. Notably there was the persistent work by the Ordnance Department to get heavier tanks than the M4 medium into service with more powerful guns. This was equally persistently opposed by AGF, who were content with M4s and 75mm guns. Thus there was a considerable delay in getting improved tanks accepted with 76mm and 90mm guns, the M26 Pershing, for example, only finally being sent to Europe when AGF were overruled by the General Staff after the Ardennes offensive by the Germans in December 1944. As events showed, however, the Ordnance Department was generally right in predicting the need for heavier guns and thicker armour years before Army Ground Forces found out the hard way in Europe in 1944. Generally speaking the US Ordnance Department did good work in 1940-45 producing a series of fine designs which, if not outstanding by German standards, proved to be adequate, and could be produced in such numbers as to sway the balance decisively in the favour of the Allies when it came to the final reckoning.

This summary cannot close without mention of the mammoth harnessing of American industry to the tank production programme. The setting up of the vast tank arsenals by major commercial companies on behalf of the US Government is recounted in detail later. It only remains to point out that the Ordnance Department acted as an agency for the co-ordination of the firms involved and in this respect were analagous to the British Department of Tank Design in dealings with commercial firms. However, Ordnance kept a firm hold of the actual design work. So great was industrial involvement with American tank production, that the Ordnance Department set up a special office in Detroit, OCO-D, specially to deal with tank design work right on the doorstep of the production facilities.

Trends and Changes

Probably the principal reason for the failure of the British to produce the tanks they really needed until too late was the constant change of policy that was evident from 1939 right through to the end of the war. Prior to 1939, British tank policy had crystallised into three distinct types: the light tank for the scouting and reconnaissance role, the cruiser tank for the exploitation role, and the infantry tank for the infantry support role. None of these classes was comparable in performance or armour protection, although the infantry and cruiser types had similar gun armament. The limited value of the light tank was clearly demonstrated in the early campaigns of the war; it was of negligible use against any but the lightest enemy tanks, but unfortunately it was frequently called upon to perform the cruiser role, for which it was entirely unsuited, mainly because there were not enough cruiser tanks due to peace-time financial restrictions, which had limited tank procurement. The light tank was thus swiftly dropped other than in the limited airborne role. Cruiser tanks and infantry tanks in 1939-41 suffered from being brought into service too quickly before the teething troubles could be overcome, and almost without exception the British tanks of these early years were dogged by mechanical shortcomings as much as by design limitations and this restricted their development potential.

4. In the beginning . . . all of nearly 150 tanks of the American 1st Armored Division lined up on manoeuvres in June 1941. At this time there were fewer than 400 tanks in American service, just about enought to equip one full armoured division on the (theoretical) 1941 scale of establishment. Note the predominance of M2 series light tanks and M1 series combat cars, with just six M2 and M2A1 medium tanks in the front row.

5. In 1940, due to limitations on pre-war procurement, British tank strength was still over-dependent on obsolete light tanks. Here Australian infantry are seen exercising with Light Tanks Mk II of the Western Desert Force in Libya.

When the need for more powerful 6pdr guns was appreciated in 1941, the guns were ready before there were tanks to take them, and when the American-built medium tanks appeared in British service in 1942 with 75mm guns, British policy changed again in favour of these weapons. However, by the time the 75mm gun was in full service in 1944 on what had then become the nearest to a British "standard" tank, the Cromwell, the Germans had produced more powerful guns and tanks, which outclassed this new vehicle. The Cromwell itself, like the earlier cruiser tanks from which it had been developed, suffered from size limitations, and among other things this prevented the fitting of the powerful 17pdr gun, which was required to match the best German tanks. Attempts to produce an enlarged version of the Cromwell, the Challenger, to take the 17pdr, were largely unsuccessful, and ironically enough it was the Sherman (M4) tank, an American vehicle, fitted with the 17pdr gun (as the Firefly), that became the most powerfully armed tank in service with the British in 1944. The Sherman had also become the most important tank in British as well as American service.

The slow moving and heavily armoured infantry tank had, with the well known exception of the Churchill, almost been discarded by 1945, again largely because this type could not be up-gunned without major redesign. In 1945, British policy was moving towards the "universal" chassis with standardised components, which had mobility close to that of the cruiser tank, and was capable of future development to mount larger calibre guns. All these lessons had been learned largely from German and American policies. The Centurion of 1945, combining the qualities of the old cruiser and infantry tanks, but with vastly superior gun power, was the successful end-product of changing British policies. It was, however, too late for the war it was designed to influence, an ironical final comment on a none too happy record of development.

Specialised vehicles

If British gun tanks left much to be desired (though tactical handling of tanks was another factor, beyond the scope of this book), the related field, in which traditional British compromise and eccentric ingenuity really excelled, was the perfection of special purpose tanks. The fighting in the Western Desert showed very early on the value of tanks capable of clearing mines, while the abortive Dieppe raid of August 1942 showed the need for a range of vehicles for the assault role in large scale operations like that planned for the future invasion of Europe. Thus the British—largely through the specially-formed 79th Armoured Division—developed a whole range of "funnies" tailor-made for mine-clearing, bridging, flame-throwing, and the assault engineer role among others. All are described in this book. Most of these were highly successful, even though they were based, in some cases, on inherently unsuitable basic chassis.

In the United States the overall picture of tank development in World War II was a much happier one. The Ordnance Department was more firmly entrenched, and accordingly more influential and autonomous, when it came to tank design, than their British opposite numbers. There was much less designing "by committee", and the United States had the added advantage of being able to stand to one side and watch events in the crucial years 1939-41 so that the Ordnance Department was able to learn the lessons demonstrated in the European and Desert campaigns. Though official American tank development between the wars had been much less dynamic than development in Britain and lagged behind in doctrine, the vast resources available in industry allowed the Ordnance Department to expedite a tank programme on an enormous scale with designs (like the M3 and M4), which were basically simple from a manufacturing point of view, unsophisticated from an operating point of view, yet capable of future development and modification. Above all they were designed with a big gun from the start, so that it was the appearance of American-built tanks in British service in 1942 that finally helped compensate for the deficiencies in the British vehicles then available.

It was largely due to the far-sightedness of the US Ordnance Department in 1942 that a new generation of tanks was available for production in late 1944, when the M4 medium was finally shown to be out-moded. From the moment the M4 entered production, Ordnance worked to improve the breed, the T20 range of medium tanks, culminating in the T26 (which evolved into the M26 Pershing), standing as evidence of the effort expended. As mentioned above, however, attempts to get Army Ground Forces to accept, or even see, the need for heavier tanks and heavier guns, was the hardest task of all for the Ordnance Department in World War II. It took more than a year to gain acceptance for the 76mm gun as a replacement for the 75mm gun in M4 medium tanks, and as long again to get the M26 into service with its 90mm gun, although both could have been in production many months earlier than they were. Armored Force doctrine saw the medium tank as a weapon of exploitation avoiding conflict with enemy tanks whenever possible and going for "soft" targets instead. In fact a stock excuse in AGF for rejecting the 90mm gun was that such a weapon in tanks would encourage crews to stalk enemy tanks instead of leaving them to the tank destroyers (where—it is fair to add—the 90mm gun was accepted much earlier). However, with German tanks, like the Tiger, mounting 88mm guns it was inevitable that an American tank to match such a vehicle would eventually be needed. Fortunately the M26 Pershing was available in time, due to Ordnance Department persistence.

"Combat Teams"

While Britain discarded light tanks, the United States continued development of this class of vehicle. It became clear from observation of the Western Desert fighting of 1941-42 (where the British were using M3 series light tanks) that even these vehicles needed a bigger gun and heavier armour. Thus, while the British concentrated on armoured cars for the reconnaisance role in place of light tanks, the Americans went on to produce a new generation of light tanks, the M24 Chaffee series, which had armour and hitting power superior to the medium tanks with which America started the war. The M24 series also started a rationalisation programme in American tank design whereby the "second generation" of wartime AFVs was to be produced in three so-called "combat teams", which each formed a series of related vehicles utilising common drive and chassis components. A "Light Weight Combat Team" was based on the M24, a "Medium Weight Combat Team" was based on the M4A3 medium tank, and a "Heavy Weight Combat Team" was based on the M26. In each case there was to be a basic tank, a recovery vehicle, a gun motor carriage, a howitzer motor carriage, and an AA gun motor carriage on the common chassis, the idea being to reduce production, maintenance, and stores-carrying problems to a minimum by making use of a maximum of standardisation. In fact, the war ended before these aims were fully realised.

In the field of special purpose vehicles, the Americans

6. Ferrying an M2 medium tank of the American 1st Armored Division across a river during exercises in June 1941.

7. In the United States almost all pre-war tanks had been built by Rock Island Arsenal. To produce tanks on the massive new scale demanded by rearmament, industry was brought in. The first big new tank plant was the Detroit Arsenal, built (in six months) and managed by Chrysler, initially to turn out M3 medium tanks, seen here in July 1941.

were no less inventive than the British. Aside from flame-throwers, however, few of the types developed by them were good enough to be accepted for service, most American mine-clearing equipment for example being clumsy and impractical. Additionally some influential American field commanders were opposed to the wholesale adoption of special purpose armoured vehicles. Thus in 1944, for the invasion of Europe, the US Army used Flail and DD equipment, which had been developed in Britain for American-built M4 mediums. British influence was also apparent in American post-war bridgelayer and assault tank designs. Other common British tank features which found their way into later American designs included rear instead of front drive, wireless installation in the turret instead of the hull, and turret-mounted bomb throwers.

The two British Commonwealth countries, Australia and Canada, which actually got as far as producing tanks of their own in the war years, have always received less than their fair share of recognition for the tremendous effort and initiative involved. In each case their indigenous design started off based on the American M3 Medium tank, the contemporary vehicle which provided the obvious and best choice at the time, 1940-41. Both the Canadian and Australian vehicles ended up vastly superior to the original M3, the Australian Sentinel, in fact, ultimately becoming different in all respects for reasons outlined in the relevant development history. Neither of these Commonwealth tanks saw combat service as such, because by the time they reached production status, American tank output had reached such a mammoth scale that the United States could supply virtually all Allied needs. In the event the Canadian tank, the Ram, still managed to play a most important part in the war, but as a special purpose vehicle in several essential but unglamorous roles.

Total American tank production in 1940-45 amounted to the vast figure of 88,410 vehicles, while in Britain over the same period, 24,803 vehicles were produced. The main enemy, Germany, turned out 24,360 tanks in World War II.

8. One of the more potent special purpose developments was the flame-thrower. This is the American M4A2 medium tank with M3-4-3 flame-throwing device.

VEHICLE NOMENCLATURE

(1) British

IN the period covered by this book all tank designs originating from War Office specifications or requirements were allotted an ordnance designation, at an early stage, in the "A" series. An example is A33, the numbers being allocated chronologically. A full list of these is given in Appendix 3. Development sub-variants were identified by an "E" number added to the main designation in the form of a suffix, an example being A7E2. Exceptions to the above were such vehicles as the Valentine and TOG which originated from agencies outside the War Office and never received "A" numbers. Occasionally, also, other suffixes were used to indicate design sub-variants, a prime example being A22F, the Churchill VII.

By the time prototype or late design status had been reached, the vehicle had normally been given an army designation indicative of its type and function. Thus the A11 became the Infantry Tank Mk I, the A12 was the Infantry Tank Mk II, the A9 became the Cruiser Tank Mk I, and the A15 was the Cruiser Tank Mk VI, to quote a few examples. Prior to 1938, vehicles of cruiser tank size and type were known as medium tanks. In the early days before this system was fully rationalised the "A" number was sometimes used as part of the designation. Thus the A13 series were first known as Cruiser Tank A13 Mk I, Mk II, and Mk III, and later as Cruiser Tank Mk III, IV, and V respectively. Further modifications to the basic model might be indicated by a letter, eg, Light Tank Mk VIC, and additional or minor modifications by one or more stars (*). An example: Medium Tank Mk II*. In most cases the suffix letter indicated a chassis or structural change and the star an armament change, but there were exceptions to this general rule. With the proliferation of new designs in 1940, the War Office started the practice of allocating type names to make designations easier to remember. Thus the Cruiser Tank A13 Mk III, which had become the Cruiser Tank Mk V was named Covenanter. It was now fully designated Cruiser Tank Mk V, Covenanter. Sub-variants under this system were indicated by mark numbers after the name. Thus, Infantry Tank Mk IV, Churchill III. This became very confusing due to the large numbers of variants and types being developed, and from early 1943 the type numbers were dropped, so that the vehicle mentioned above would become Infantry Tank, Churchill III (or Churchill Mk III).

In this book we have generally used the *simplest* designation appropriate to the vehicle concerned and in sections describing variants we have omitted repetition of the type designation for simplicity. Thus we refer to the Covenanter I, for example, rather than "Cruiser Tank Mk V, Covenanter I". Close-support tanks, which were armed with howitzers instead of guns, were indicated by the initial CS.

Self-propelled guns were designated so as to show the basic chassis on which they were mounted, and calibre of the weapon. In the earlier part of the war the chassis was known as a "carrier" for the gun. Thus the 25pdr on the Valentine chassis was known as the "Carrier, Valentine, 25pdr gun Mk I, Bishop". The mark number referred not to the gun but to the chassis. Later the term SP (self-propelled) was used and the 17pdr gun on the Valentine chassis, produced in 1944, was the "SP 17pdr, Valentine, Mk I, Archer". In this case the mark number referred to the item of equipment as a whole, thus coming into line with the mark numbers allotted to tanks. Other special purpose types were usually designated in a similar way with the mark number referring to the equipment as a whole, and not the basic tank chassis on which it was built. Thus the Sherman ARV Mk I and the Sherman ARV Mk II indicated different types of recovery vehicle, though each was built on the Sherman V chassis.

Finally, we must point out that while a Cruiser Tank was

colloquially described as such, on British ordnance inventories it would be described as "Tank, Cruiser, etc, etc". We have used this description where appropriate in the data sections, but elsewhere have used the less formal description.

(2) United States

IN America a much more rational designation system was used. In the project, design, and development stages, a vehicle was given a designation in the T series (best remembered as T for Test). Thus a vehicle might be designated T89. Any experimental modification was indicated by a suffix in the E series (E for Experimental). Thus T1E1, T25E1, or T20E3, in the latter case the "3" indicating the third experimental modification. "T" numbers were normally allocated chronologically. When fully accepted for service by the using arms, the vehicle was "standardised" and given a designation in the M series. Thus M6 or M8. It was not usual for the M number to bear any relation to the original T designation, but towards the end of the war there was a change in favour of this in an attempt to avoid confusion. Thus the Light Tank T24 became the M24 on standardisation, for example. In rare instances a design was standardised from the "drawing board" and never received a T designation; an example was the Medium Tank M3. There were also many instances where vehicles were put into limited production and service before being standardised, and in some cases they never achieved the status of standardisation —an example being the T23 medium tank.

At this stage it must be emphasised that this system of designation was used for every item of military equipment in the US Army, so that it was possible to have an M3 Medium tank, an M3 Light tank, an M3 gun mount, an M3 rifle, an M3 flame-gun, an M3 gun sight and so on. Thus it was normal practice to qualify each item by its full title. Strictly speaking, therefore, it is necessary to say Light Tank M3 to distinguish it from Medium Tank M3 and so on. Within the context of this book, however, we have omitted repetition of the qualifying description for brevity's sake where the meaning is apparent from the text.

Any modification to a basic vehicle after standardisation was indicated by an "A" suffix. An example of this is seen clearly in the Medium Tank M4 series, where engine and other changes gave rise to the M4A1, M4A2, M4A3 etc. Modifications confined to the chassis only were indicated by a "B" series suffix. An example arises in M7 howitzer motor carriage development. The M7 was based on the M3 medium tank chassis, and the same design based on the M4A3 chassis became the M7B1. Had yet another chassis been used subsequently, the designation would have been M7B2, and so on. The "E" series suffix was rarely retained when a design was standardised, but there were exceptions, one such being Assault Tank M4A3E2. It should be borne in mind that any special purpose equipment carried on American tanks was designated separately but following the same system. Thus an M4A1 tank could be seen fitted with an M1 dozer blade or a T34 rocket launcher, etc.

Self-propelled artillery in American service was described by the calibre of the weapon with the term "gun/howitzer/mortar motor carriage" as appropriate. Example: 105mm Howitzer Motor Carriage M37. Other special purpose vehicles were designated similarly to tanks. Example: Tank Recovery Vehicle T1. Names were not officially used for American tanks until the M26 heavy tank was called the Pershing. Before that, however, British names for American equipment (eg, Sherman, Lee) were being used colloquially in American service and some American vehicles had unofficial but commonly used names, such as Hellcat for the M18 GMC or Jumbo for the M4A3E2 assault tank.

Instead of being classified as "standard" equipment, American AFVs were sometimes classified "limited standard". This category was given to a vehicle which was not fully satisfactory for universal service, but which could be used when necessary. A further classification was "substitute standard", usually given to obsolescent or expedient equipment due for early replacement but which could still be used pending availability of the new design. Finally there was the "limited procurement" classification given to vehicles for which only restricted use could be foreseen. As the term implies, such vehicles were usually produced only in small quantities. Classification could, of course, be changed as necessary for any given vehicle type.

9. In its final form the M4 Medium tank was a potent weapon with 76mm gun and many improvements over earlier models. It was still too thinly armoured however. This 7th Army M4A3 (76mm) HVSS has added sandbag protection, a typical extemporised improvement. March 1945.

ABBREVIATIONS AND TERMS USED IN THIS BOOK

AA	anti-aircraft
AGF	Army Ground Forces
Amd Div	Armoured Division
ACF	American Car and Foundry Co
AFV	armoured fighting vehicle
American Loco	American Locomotive Co
AP	armour piercing
APCBC	armour piercing capped, ballistic capped
APC	armoured personnel carrier or armour piercing (capped)
APG	Aberdeen Proving Ground, Maryland
ARV	armoured recovery vehicle (British term)
ASF	Army Service Forces
Baldwin	Baldwin Locomotive Works
BARV	beach armoured recovery vehicle (British term)
Birmingham Carriage & Wagon	Birmingham Railway Carriage and Wagon Co
Cadillac	Cadillac Division of General Motors
cal	calibre (in inches or millimetres as appropriate)
Canadian Pacific	Canadian Pacific Railway Co
CDL	canal defence light
Co	company
Corp	corporation
CS	close support
Detroit Arsenal	Detroit Tank Arsenal, Center Line, Michigan
eg	for example
Fisher	Fisher Division of General Motors
Federal Welder	Federal Welder & Machine Co
Ford	Ford Motor Co, Detroit
ft	feet
FVRDE	Fighting Vehicle Research and Development Establishment (British term)
GM or GMC	General Motors Corporation
GMC	gun motor carriage (American term)
Grand Blanc	Grand Blanc Arsenal, Michigan
HE	high explosive
HEAT	high explosive anti-tank
HMC	howitzer motor carriage (American term)
hp (or HP)	nominal brake horse power
HQ	headquarters
ie	that is
in	inches
LAD	light aid detachment (British term)
Lima	Lima Locomotive Co
LMS	London Midland & Scottish Railway
LVT (A)	landing vehicle tracked (armoured)
Marmon-Herrington	Marmon-Herrington Co
MEXE	Military Engineering Experimental Establishment (British term)
MG	machine gun
Mk	Mark

mm, cm	millimetres, centimetres
MMC	mortar motor carriage (American term)
Montreal	Montreal Locomotive Works
mph	miles per hour
MV	muzzle velocity
OCO-D	Office, Chief of Ordnance, Detroit
OP	observation post
OQF	ordnance quick firing
Pacific Car & Foundry	Pacific Car and Foundry Co
pdr	pounder (gun calibre)
Pressed Steel	Pressed Steel Car Co
Pullman	Pullman Standard Car Co
QMC	Quartermaster Corps
(qv)	(which see)—cross reference indicator
RA	Royal Artillery
RE	Royal Engineers
REME	Royal Electrical & Mechanical Engineers
RHA	Royal Horse Artillery
Rock Island	Rock Island Arsenal
ROF	Royal Ordnance Factory (Woolwich)
rpm	revolutions (or rounds) per minute
SP	self-propelled (British term)
TRV	tank recovery vehicle (American term)
Vickers	Vickers-Armstrong Ltd (Chertsey, Surrey, and Newcastle-upon-Tyne)
Vulcan	Vulcan Foundry Ltd
WD	War Department
USMC	United States Marine Corps
— (in data section)	figure not known or unrecorded

PART 1

BRITISH VEHICLES

10. A Covenanter tank in Perth, Scotland, in 1941, parading in a recruiting tour for the Royal Armoured Corps. Though used widely for training, the Covenanter was not used in action.

11. Light Tank Mk II Indian Pattern.

THE Light Tank Mk II was a production development of the "private venture" Carden-Loyd Mk VII and VIII tanks, which were supplied to the War Office A4 specification in 1929; these Carden-Loyd vehicles were built by Vickers (who absorbed the Carden-Loyd company in 1928) and were designated Light Tank Mk I by the British Army. Five vehicles were used for trials and various suspensions were tested. The Mk II, also built by Vickers, had a similar hull to the Mk I, but a larger turret and Rolls-Royce engine. Horstmann-type suspension was standard with horizontal coil springs (known as the "two pair" type). The Mks IIA and IIB had detail improvements, including better engine cooling and re-sited fuel tanks. The Mk III was a slightly modified version of the Mk II, built by the Royal Ordnance Factory at Woolwich and incorporating improvements suggested by experience. These included a higher roomier hull and the later type of Horstmann suspension with angled coil springs giving a longer "throw". These were known as "four pair" bogies. Only 36 of these tanks were built. The other variant in this early series of light tanks was the Mk II Indian Pattern, built for the Indian Army, which differed mainly in having a cupola on the turret. Production of the Mks II and III was completed in 1936 and these types were mainly replaced by later marks by 1939-40. Remaining vehicles were mostly used for training and instruction until 1942, but a few Mk IIA and IIB remained in service with tank battalions of the Western Desert Force in 1940. Mk IIAs and Mk IIIs were also used by a South African battalion in the Abyssinian campaign of 1941.

The Mk II-III series were two-man vehicles (driver and commander) with side-mounted engines and of light riveted construction. They were of negligible tactical value under World War II conditions.

SPECIFICATION

Designation: Tank, Light, Mk II, or IIB; Tank, Light, Mk III.
Crew: 2. Duties: Driver/Commander.
Battle weight: 9,520 lb (Mk III, 10,080lb).
Dimensions: Length 11ft 9in (Mk III, 12ft).
Height 6ft 7½in (Mk III 6ft 11in).
Width 6ft 3½in.
Track width 9½in
Track centres/tread 5ft 2½in
Armament: 1 × Vickers ·303 MG
(1 × Vickers ·50 MG in some Mk IIIs)
Armour thickness: Maximum 10mm (Mk III, 12mm).
Minimum 4mm.
Traverse: 360°. Elevation limits: +37° to −11°
Engine: Rolls-Royce 6 cylinder 66hp
Maximum speed: 30mph
Maximum cross-country speed: 20 mph (approx)
Suspension type: Horstmann coil spring
Road radius: 130 miles
Fording depth: 2ft 3in
Vertical obstacle: —
Trench crossing: —
Ammunition stowage: —
Special features/remarks: Many Mk IIA and 11B were fitted retrospectively with "four pair" bogies as in the Mk III.

12, Tank, Light Mk IIA shown in use for crew training, 1940. Note the "two pair" spring bogies.

13. Light Tank Mk IIB fitted with "four pair" bogies, Western Desert 1940. Mk III similar.

LIGHT TANK, VICKERS COMMERCIAL TYPE

United Kingdom

FORTY Vickers light tanks (Model 1936) on order for the Netherlands Army were taken over by the War Office on completion in 1939 and were used for training by the British Army for the duration of the war. None was used operationally. They were known as "Dutchmen" to the British, though this was an unofficial designation.

Vickers were active throughout the thirties selling their own designs to overseas governments. The "Dutchmen" were mechanically similar to the Vickers Mk IV light tank and had a Meadows 88hp engine and a simpler hull shape with hexagonal turret. Armament and other details were as for the Mk II.

14. A Vickers commercial type ("Dutchman") in use for training at the School of Military Engineering, 1943.

15. Close view of a "Dutchman" showing its distinctive hexagonal turret.

LIGHT TANK, Mks IV and V

THE Light Tank Mk IV was a more powerful development of the Mk III and reverted to the Meadows engine as used in the Light Mk I. In the Mk IV the suspension was further modified to dispense with the idler wheel, the bogies being respaced accordingly. The hull was re-shaped at the rear to give more room internally. In other respects it was similar to the Mk II and III. As with the Mk II, there was also an Indian Pattern version with a cupola on the turret.

The Light Mk V, which first entered service in 1935, incorporated several improvements intended to overcome the deficiencies of the earlier models. The hull was lengthened slightly to allow the fitting of a two-man turret, and two co-axial machine guns were fitted. The larger turret incorporated a commander's cupola and the heavier weight aft of both the turret and the extra crew member greatly improved the handling qualities; a return roller was added to the front suspension bogies.

Both the Mk IV and Mk V were obsolete at the start of World War II, though a few remained in front-line service with units equipped with later marks. Most were employed for training, however, in the various tank schools in Britain and overseas.

VARIANTS

Two light Mk V chassis were used for experiments with AA mounts in 1940. One was fitted with a twin 15mm Besa machine gun mount in place of its turret, and the other had a Boulton & Paul quadruple Browning power-operated aircraft turret fitted. At first this retained the Perspex canopy, later removed. Light armour folding flaps were fitted for crew protection. These were development vehicles only.

16. A Light Mk IV used for training on Salisbury Plain, 1940. Note the hand grips for the commander.

SPECIFICATION

Designation: Tank, Light, Mk IV
Crew: 2 (commander, driver)
Weight: 9,520lb
Dimensions: Length 11ft 2in; Height 6ft 8½in; Width 6ft 11½in; Track width 9½in; Track centres/tread 5ft 8½in
Armament: Main: 1 × Vickers ·303 MG or 1 × Vickers ·5 MG
Armour thickness: Maximum 12mm; Minimum 4mm
Traverse: 360°. Elevation limits: +37° to −10°
Engine: Meadows 6 cylinder 88hp
Maximum speed: 36mph
Maximum cross-country speed: 28mph (approx)
Suspension type: Horstmann coil-spring
Road radius: 130 miles
Fording depth: 2ft
Vertical obstacle: —
Trench crossing: —
Ammunition stowage: —
Special features/remarks: Indian Pattern had cupola on turret.

17. Light Tank Mk V with twin 15mm Besa guns for AA role.

SPECIFICATION

Designation: Tank, Light, Mk V
Crew: 3 (commander, gunner, driver)
Weight: 10.740lb
Dimensions: Length 12ft 10 in Track width 9½in
Height 7ft 3in Track centres/tread 5ft 8½in
Width 6ft 9in
Armament: Main: 1 × Vickers ·303 MG
and 1 × Vickers ·5 MG
Armour thickness: Maximum 12mm
Minimum 4mm
Traverse: 360°. Elevation limits: +37° to −10°
Engine: Meadows 6 cylinder 88hp
Maximum speed: 32.5mph
Maximum cross-country speed: 25mph (approx)
Suspension type: Horstmann coil-spring
Road radius: 130 miles
Fording depth: 2ft
Vertical obstacle: —
Trench crossing: —
Ammunition stowage: —

18. Light Mk V fitted with quad Boulton & Paul aircraft-type turret. Note folding side armour plates.

19. Light Tank Mk V of the 9th Lancers.

20. Light Tank Mk VIA.

FOLLOWING on from the Mk V light tanks, the Mk VI was identical in all respects except for the turret which was further redesigned to give room in the rear for a wireless set. In the Mk VIA the single return roller was removed from the top of the leading bogie and attached to the hull sides. The Mk VIB was a vehicle mechanically similar to the VIA, but with detail differences to simplify production. These alterations included a one-piece armoured louvre over the radiator (instead of a two-piece louvre) and a plain circular cupola in place of the faceted type of the VIA. The Light Mk VIB Indian Pattern, produced for the Indian Army, was identical to the standard model except for the removal of the commander's cupola in favour of a plain hatch in the turret roof. The Mk VIC, last of the series, also lacked the commander's cupola and was more powerfully armed than its predecessors, having co-axial 15mm and 7·92mm Besa machine guns in place of the ·303 and ·5 Vickers machine guns of the earlier marks. It also had wider bogies and three carburettors in the engine to give improved performance.

The Mk VI series entered production in 1936 and VIC production tailed off in 1940. These were in wide service with the British Army at the outbreak of war in 1939, the VIB being produced and used in the largest numbers. The bulk of British tank strength in 1940, in France, in the Western Desert, and elsewhere was in fact composed of Mk VI lights and they were frequently called on to act in the "cruiser tank" role rather than the reconnaisance role for which they were intended—usually with heavy losses to themselves. After Dunkirk, these light tanks were also widely used to equip armoured divisions in Britain and were still in quite wide "first line" use as late as 1942, though by then being replaced by later types and relegated to training.

VARIANTS

Tank, Light AA Mk I: Experience under the *Blitzkrieg* conditions of the German invasion of France and Flanders in May 1940, when the British first encountered co-ordinated air strikes in support of attacks by enemy armoured divisions, led to the hasty development of specialist AA tanks. Experimental versions, based on the Mk V chassis, have

21. Light Tank Mk VIB Indian Pattern. Note absence of cupola.

22. Tank, Light AA Mk I.

23. Tank, Light AA Mk II.

already been described. Production vehicles had a power-operated turret with four 7·92mm Besa MGs set in a modified superstructure. The first variant, Tank, Light AA Mk I, was on a Mk VIA chassis.

Tank, Light AA Mk II: This was essentially the same as the Light AA Mk I, but had improvements in the form of better sights and a roomier, more accessible, turret. It also had an external ammunition bin mounted on the hull rear. The Tank, Light AA Mk II was built on the Light Mk VIB chassis. A troop of four AA tanks was attached to each regimental HQ squadron.

Tank, Light, Mk VIB, with modified suspension: A small number of Mk VIBs were fitted with larger diameter sprocket wheels and separate idlers at the rear (as in the Light Mk II) in order to give longer ground contact and improved ride. This modification was only experimental, however, and was not adopted as standard.

Tank, Light, Mk VI, Bridgelayer: One chassis was converted by MEXE in 1941 to carry a light folding scissors bridge. Shipped to Middle East Forces for combat trials, it was apparently lost en route and no further record of it exists.

Specification

Designation: Tank, Light, Mk VI, VIA, VIB, VIC
Crew: 3 (commander, gunner, driver)
Weight: 11,740lb (Mk VI, 10,800 lb)
Dimensions: Length 12ft 11½in; Track width 9½in
Height 7ft 3½in (Mk VIC 7ft); Track centres/tread 5ft 8½in
Width 6ft 9in
Armament: Main: 1 × Vickers ·5 MG (Mk VIC, 1 × Besa 15mm MG)
Secondary: 1 × Vickers ·303 MG (Mk VIC, 1 × Besa 7·92mm MG)
Armour thickness: Maximum 14mm
Minimum 4mm
Traverse: 360°. Elevation limits: +37° to −10°
Engine: Meadows 6 cylinder 88HP
Maximum speed: 35mph
Maximum cross-country speed: 25mph
Suspension type: Horstmann coil-spring
Road radius: 130 miles
Fording depth: 2ft
Vertical obstacle:—
Trench crossing: —
Ammunition stowage: —
Special features/remarks: Final development of Vickers Carden-Loyd light tank series.

24. Light Tank Mk VIB in use for training, 1940. Note the circular cupola and single radiator louvre.

25. Light Tank Mk VIC of 8th Armoured Division, 1940.

26. Light Tank Mk VIB with modified suspension.

27. Standard production Tetrarch I.

FOLLOWING the Light Tank Mk VI, Vickers designed a larger type of light tank in 1937 capable of mounting a 15mm Besa gun in place of the machine guns which were mounted in previous Vickers light tanks. For the Mk VII Vickers adopted a novel type of suspension incorporating large road wheels. As with some earlier Vickers designs, however, steering was achieved by flexing the tracks. The Mk VII was offered to the War Office in 1938 (when it was known unofficially as the "Purdah") and the vehicle, designated A17, was tested for possible future adoption as a "light cruiser" tank, since War Office light tank requirements were already being met by Mk VI production. An order for 120 vehicles was placed at the end of 1938, when war seemed inevitable, and Vickers were asked to transfer production to Metropolitan-Cammell to be free to do other work. The order was subsequently doubled, but production was delayed, partly by the transfer, partly by a War Office decision to concentrate on cruiser and infantry tanks following the Dunkirk evacuation, and partly by bombing of the factory. Finally only 177 vehicles were produced, the first being completed in November 1940. Some were issued to armoured divisions in Britain at this time since there was a shortage of heavier tanks. War Office policy had by this time changed in favour of armoured cars, rather than light tanks, for reconnaissance work. Consideration was given to sending the Light Mk VIIs to the 8th Army, but inadequate cooling arrangements made this type unsuitable for desert use. A small number were subsequently sent to the Soviet Army under Lend-Lease, and one squadron of Light Mk VIIs was used in the invasion of Madagascar in May 1942, the first time the type was used operationally.

Meanwhile, the development of airborne forces suggested the adoption of the Light Mk VII for the glider-borne role. The name Tetrarch was adopted in 1943 for the Light Mk VII, and the Hamilcar glider was specially designed to carry one of these vehicles. An airborne reconnaissance regiment was specially formed as part of 6th Airborne Division for the invasion of Europe. One squadron of Tetrarchs and Universal Carriers was carried into action on June 6, 1944, the first day of the "Overlord" landings by gliders, though their contribution to the success of the landing was limited, partly because their number was so small.

A few Tetrarchs remained in use until about 1949, when the British abandoned gliders and the Tetrarchs were also withdrawn.

VARIANTS

Tetrarch I CS: A few vehicles were converted for the close support (CS) role, with 3in howitzers substituted for the 2pdr. A few standard Tetrarch Is were also fitted with Littlejohn adaptors on their 2pdrs to increase muzzle velocity.

Tetrarch DD: One Tetrarch was fitted and tested with a propeller drive and canvas collapsible flotation screens in June 1941. Trials of the Tetrarch as a swimming tank—chosen for its lightness—were carried out in Brent Reservoir. These proved successful and Straussler DD (Duplex Drive) conversion, whereby a propeller was driven by power take-off from the engine, was adopted for the Valentine and Sherman, the latter being used in the Normandy landings.

28. Tetrarch I CS with 3 in howitzer.

SPECIFICATION

Designation: Tank, Light Mk VII, Tetrarch I
Crew: 2 (commander, gunner, driver)
Battle weight: 16,800lb
Dimensions: Length 13ft 6in; Height 6ft 11½in; Width 7ft 7in; Track width 9½in; Track centres/tread 6ft 6in
Armament: Main: 1 × 2pdr QFSA (1 × 3in howitzer in Mk I CS)
Secondary: 1 × 7·92mm Besa MG
Armour thickness: Maximum 14mm
Minimum 4mm
Traverse: 360°. Elevation limits: —
Engine: Meadows 12 cylinder 165hp

29. Tetrarch DD on test, showing propeller and flotation screen.

Maximum speed: 40mph
Maximum cross-country speed: 28mph
Suspension type: Steerable steel road wheels, independently sprung.
Road radius: 140 miles
Fording depth: 3ft
Trench crossing: 5ft
Ammunition stowage: 50 rounds 2pdr.
2,025 rounds 7·92mm.
Special features/remarks: Littlejohn adaptor on guns of some vehicles.

LIGHT TANK, VICKERS 6 TON TYPE B — United Kingdom

DURING the thirties Vickers-Armstrongs were active in the tank design and production field, offering vehicles of their own conception to foreign nations as well as fulfilling contracts for the British War Office. Apart from the light tanks offered for overseas sale, the most prolific Vickers type was the 6 tonner which was supplied to several countries from 1929 onwards, including Russia, China, Bulgaria, Finland, Poland, Portugal, Thailand, Bolivia, Greece. Basically it was supplied as Type A with two Vickers ·303 machine guns similar to the light tanks, or Type B with a 47mm gun and one co-axial machine gun. Some vehicles had different calibre weapons to these according to requirements of purchasers. The Vickers 6 tonner was an excellent design for its period, low, simple mechanically, and well-armed and armoured for its size. The British Army never ordered this type, but on the outbreak of war in September 1939, the British took over several 6 tonners which were built for other nations, and they were used in a training role for the rest of the war. By British standards this vehicle was a light tank. Many features of the Vickers 6 tonner were copied in the American Light Tank T1, prototype for the American M1-M3 light tank series.

SPECIFICATION

Designation: Vickers 6 ton Type B
Crew: 3 (commander, gunner, driver)
Battle weight: 15,680lb
Dimensions: Length 15ft; Height 7ft 2in; Width 7ft 11in; Track width 8in; Track centres/tread 7ft
Armament: Main: 1 × 47mm QFSA
Secondary: 1 × Vickers ·303 cal MG
Armour thickness: Maximum 17mm
Minimum 5mm

30. Vickers 6 ton Type B in use with a training unit, 1940.

Traverse: 360°. Elevation limits: —
Engine: Armstrong Siddeley 4 cylinder air-cooled 80hp
Maximum speed: 22mph
Maximum cross-country speed: 14mph (approx)
Suspension type: Box bogie and leaf spring
Road radius: 100 miles
Fording depth: 3ft
Vertical obstacle: 2ft 6in
Trench crossing: 6ft
Ammunition stowage: 50 rounds 47mm
4,000 rounds ·303 cal
Special features/ remarks: Training vehicle only; guns removed on some.

31. Standard production Harry Hopkins.

ORIGINALLY known as the "Tank, Light Mk VII, revised", the Mk VIII was designed by Vickers in 1941 as an improved version of the Mk VII. Three prototypes were built, utilising chassis and mechanical components of the Mk VII but incorporating a revised faceted hull and turret to give better shot deflection. Armour maximum was increased, hydraulically assisted steering was fitted, and there were many detail refinements. Metropolitan-Cammell undertook production as a follow-on to the Mk VII, and an order for 99 vehicles was completed in 1944. Its general characteristics were the same as the Tetrarch, but the Harry Hopkins never entered service as the British light tank requirement was by 1944 limited only to the airborne role, for which stocks of Tetrarchs were adequate.

The Harry Hopkins chassis was used as the basis for the Alecto self-propelled 95mm howitzer (ie, howitzer motor carriage), which was originally designated Harry Hopkins Mk I CS. Basically this was the Harry Hopkins with turret removed, superstructure slightly modified, and a 95mm howitzer mounted low in the hull front, thus producing a fast, low, lightweight SP vehicle. The Alecto II (also known as the Alecto Recce) was a variant with a 6pdr gun replacing the howitzer. Only pilot and development vehicles were produced, however, and the cessation of hostilities terminated interest in these designs. A small number of vehicles were completed as Alecto Dozers (in 1945) with hydraulically operated dozer blades and lifting mechanism in place of the gun mount. Vickers were responsible for Alecto development, to a General Staff requirement of April 1942. The project was delayed, however, by War Office indecision as to its employment.

SPECIFICATION

Designation: Tank, Light, Mk VIII, Harry Hopkins
Crew: 3 (commander, gunner, driver)
(Alecto: 4, loader in addition to the above)
Battle weight: 19,040lb
Dimensions: Length 14ft (excluding gun in Alecto) Track width $10\frac{1}{2}$in
Height 6ft 11in (Alecto 4ft $10\frac{1}{2}$in) Track centres/tread 7ft
Width 8ft $10\frac{1}{2}$in
Armament: Main: 1 × 2pdr OQF
(Alecto 95mm or 6pdr–see text)
Secondary: 1 × 7·92mm Besa MG
(Not in Alecto)
Armour thickness: Maximum 38mm
Minimum 6mm
Traverse: 360°
(15° each side in Alecto)
Elevation limits: —
(10mm in Alecto)
Engine: Meadows 12 cylinder 148hp
Maximum speed: 30mph
Maximum cross-country speed: 20mph
Suspension type: Steerable road wheels
Road radius: 125 miles
Fording depth: 3ft
Vertical obstacle: 2ft
Trench crossing: 5ft
Ammunition stowage: 50 rounds 2pdr,
2,025 rounds 7·92mm
Special features/remarks: Littlejohn adaptor on guns of some vehicles. Alecto was basically similar except where noted.

VARIANTS

Alecto III: Project only to mount 25pdr howitzer in place of 95mm howitzer of Alecto I.

Alecto IV: Further project to mount 32pdr in place of 95mm howitzer. Neither of these two marks was actually built.

33. Alecto I with 95mm howitzer.

32. Alecto Dozer showing control box on hull top.

34. Alecto II with 6pdr gun.

MEDIUM TANK, Mk II, IIA, and II*

United Kingdom

35. Medium Mk II of a training regiment in April 1940.

THE Mediums Mk I and II, with their derivatives, were the standard Royal Tank Corps vehicles from 1923 until about 1938, when the new designs of cruiser and infantry tanks came into service. Built by Vickers they started to replace the old World War I vintage tanks in 1923 and deliveries were completed by 1928. The Medium Mk II was completely obsolete by the start of World War II, but was used for training in the first half of the war. In 1940, in addition, some were among the assorted collection of vehicles which made up the strength of armoured units after the loss of most of Britain's first line tanks in the withdrawal from France. A few other Mk IIs were used by the Western Desert Forces in 1940-41 at Mersa Matruh and Tobruk, essentially training vehicles from Egypt pressed into service to make up first line strength. The Medium Mk II was of riveted construction, and had an air-cooled engine. It was mechanically unsophisticated by World War II standards, suffered from short track life and was inadequately armoured. The training vehicle shown has its armoured side skirts removed. Mks IIA and II* incorporated various modifications, including a co-axial MG in the latter.

SPECIFICATION

Designation: Tank, Medium, Mk II
Crew: 5 (commander, driver, wireless operator, gunners (2))
Battle weight: 30,240lb
Dimensions: Length 17ft 6in; Height 9ft 10½in; Width 9ft 1½in; Track width 13in; Track centres/tread 8ft 6in
Armament: Main: 1 × 3pdr QFSA
Secondary: 3 × Vickers ·303 cal MG
Armour thickness: Maximum 8mm
Minimum 8mm
Traverse: 360°. Elevation limits: —
Engine: Armstrong Siddeley 90hp
Maximum speed: 18mph
Maximum cross-country speed: 10mph
Suspension type: Box bogie ("Japanese Type")
Road radius: 120 miles
Fording depth:—
Vertical obstacle:—
Trench crossing: 6ft 6in
Ammunition stowage: —
Special features/remarks: Armament removed from some vehicles in training units.

CRUISER TANK, Mk I and Mk ICS (A9) — United Kingdom

36. Standard production Cruiser Tank Mk I.

37. Cruiser Tank Mk ICS of 1st Armoured Division, 1940.

THE main British tank strength in the twenties and early thirties was the Vickers Medium Mk II with the various light tank types introduced for the "scouting" role. Proposed replacements for the Medium Mk II, the Medium Mk III ("16 tonner") and the "Independent" were abandoned on the grounds of expense during the financial cut-backs of the thirties. Similarly the A7 and A8, built by the Royal Ordnance Factory, Woolwich, as medium tanks, never went beyond prototype stage, again largely for financial reasons. In 1934, however, Sir John Carden of Vickers-Armstrong (the firm which built the Medium Mk III), designed a new medium tank, designated A9, to meet General Staff requirements resulting from the proposals offered to the General Staff's Research Committee by the Inspector-General of the Royal Tank Corps. It incorporated the best features of the discontinued Medium Mk III, but was much lighter so that it could be powered by a standard commercially-made engine and thus be produced more cheaply. Designed weight was about 10 tons, though production vehicles exceeded this. The pilot model was to be powered by a single Rolls-Royce Phantom II engine of 7·67 litres, but this proved unable to provide the specified performance, so a 9·64 litres AEC bus engine was adopted instead.

An alternative 2pdr gun or 3·7in howitzer (CS) armament could be fitted and there were two auxiliary machine gun turrets as in the Medium Mk III. A 3pdr instead of 2pdr was initially proposed, but the latter weapon had become the new standard tank gun when production started in 1937. Two types of tank "cruiser" (essentially the old "medium" class) and "infantry" had been decided upon by the British War Office when considering future requirements in 1936. The A9, which originally had been rated a "medium" tank, thus became the Cruiser Tank Mk I. Trials of the pilot model started in July 1936 and production of 125 vehicles commenced a year later, 50 of them built by Vickers and 75 by Harland and Wolff, Belfast. A9s equipped some regiments of the 1st Armoured Division in France until the time of the Dunkirk withdrawal in June 1940. They were also used by regiments in the Western Desert until 1941. The A9 had inadequate armour and too low a speed for the "cruiser" role. Interesting design features were the external steering brakes on the rear sprockets (good for cooling), power turret traverse, and "slow motion" suspension—later used essentially unchanged on the Valentine.

SPECIFICATION

Designation: Tank, Cruiser, Mk I (A9)
Crew: 6 (commander, gunner, loader, driver, MG gunner (2))
Battle weight: 28,728lb
Dimensions: Length 19ft; Height 8ft 8½in; Width 8ft 2¼in; Track width: 14in; Track centres/tread 7ft 3in
Armament: Main: 1 × 2pdr OQF
(1 × 3·7in howitzer in Mk ICS)
Secondary: 3 × Vickers ·303 cal MG (one co-axial)
Armour thickness: Maximum 14mm
Minimum 6mm
Traverse: 360° Elevation limits:—
Engine: AEC Type A179 6 cylinder gasoline (petrol) 150hp
Maximum speed: 25mph
Maximum cross-country speed: 15mph (approx)
Suspension type: Triple-wheel bogies on springs with Newton hydraulic shock absorbers ("Slow motion" type)
Road radius: 150 miles
Fording depth:—
Vertical obstacle: 3ft
Trench crossing: 8ft
Ammunition stowage: 100 rounds 2pdr
3,000 rounds ·303 cal MG
Special features/remarks: First British tank with hydraulic power traverse. Boat-shaped hull offering no external vertical faces. Riveted construction.

38. Cruiser Tank Mk IIA. (Production prototype vehicle A10E1).

AFTER Sir John Carden had completed the design of the A9 in 1934, the War Office asked him to produce a more heavily armoured version better able to work with the infantry. The A10, which this heavier version was designated, was mechanically and structurally similar to the A9 which it closely resembled. The requirement called for a lower speed but an armour basis of 24mm. The mild steel pilot model of the A10 was completed in July 1937 and delivered to the Army for trials. Among the changes called for as a result of these tests were alterations to the gear ratios to give higher speed, an increase in armour to 30mm to match revised requirements and the addition of a machine gun alongside the driver in the hull front.

The A10 had the same turret and the same basic boat-shaped hull as the A9, and the additional armour called for was achieved simply by bolting extra armour plates to the outside of the hull and turret structures. It was the first British tank with this type of composite construction. Hydraulic power traverse was provided for the turret, as in the A9, but the auxiliary machine gun turrets were discarded. By early 1938 the A10 was, in fact, considered too lightly armoured to fall into the revised "infantry tank" classification—even though it had been designed in the first place to work with the infantry. Therefore it was redesignated as a "heavy cruiser" tank. Production contracts were placed in July 1938 for 100 vehicles—10 by Vickers, and 45 each by the Birmingham Railway Carriage Co and Metropolitan-Cammell. Another 75 were ordered from Birmingham Railway Carriage Co in September 1939. Orders were completed by September 1940. A10s were issued to units of the 1st Armoured Division and were used, with A9s, in France in 1940 and in the Western Desert until late 1941. Orders for both the A9 and A10 were restricted since in 1936-37 it was decided to concentrate on developing new, faster, designs with Christie suspension and better armour. The A9 and A10 then became regarded as stop-gap types only. For the circumstances leading to this, see the development history of the A13 series vehicles.

VARIANTS

Tank, Cruiser, Mk II: Vehicle with same 2pdr gun mount and co-axial Vickers machine gun as in the A9. Besa machine gun fitted in 1940 in the hull front. Only 13 were built; also known as the A10 Mk I.

Tank, Cruiser, Mk IIA: Vehicle with 2pdr gun in redesigned mount, one Besa gun in co-axial position in turret, and second Besa machine gun in hull front next to driver's position. Remainder of production vehicles were of this type. Also known as A10 Mk IA.

Tank, Cruiser, Mk IIA CS: Close support variant with 3·7in howitzer replacing the 2pdr gun. Otherwise as Mk IIA. Only 30 of this type were built. For turret appearance see picture of Cruiser Mk ICS. Also known as A10 Mk IA CS.

SPECIFICATION

Designation: Tank, Cruiser, Mk II, IIA or IIA CS (A10)
Crew: 5 (commander, loader, gunner, driver, hull machine gunner)
Battle weight: 31,696lb
Dimensions: Length 18ft 4in; Height 8ft 8½in; Width 8ft 3½in; Track width 14in; Track centres/tread 7ft 3in
Armament: Main: 1 × 2pdr QFSA (1 × 3·7in howitzer in CS model)
Secondary: 2 × Besa MG
Armour thickness: Maximum 30mm
Minimum 6mm
Traverse: 360°. Elevation limits: —
Engine: As A9 (Cruiser Tank Mk I)
Maximum speed: 16mph
Maximum cross-country speed: 8mph
Suspension type: As A9
Road radius: 100 miles
Fording depth: 3ft
Vertical obstacle: 2ft
Trench crossing: 6ft
Ammunition stowage: 100 rounds 2pdr, 4,050 rounds ·303 cal or 7·92 cal MG
Special features/remarks: As A9. First British tank with composite armour construction. Mk IIA was first British tank to be armed with the Besa MG.

39. Standard production Cruiser Mk III.

THE AI3 was an important step in the development of British tanks since it was the design which initiated the long run of Cruiser tanks with Christie suspension produced in the World War II period by the British. Essentially it stemmed from the designs developed (largely unsuccessfully in his own country) by the American designer J. Walter Christie. Responsible for the introduction of the Christie suspension into British tanks was Lt Col G. Le Q. Martel, one of the pioneers of British tank development in the twenties, who was appointed Assistant Director of Mechanisation at the War Office in late 1936, and as such was in charge of AFV development. In September 1936, soon after appointment, Martel attended the Soviet Army autumn manoeuvres and was much impressed by the speed and performance of the BT tank, which the Russians had developed and put in service in large numbers after buying some of Christie's prototypes. Returning to London, Martel expressed the opinion that a tank of vastly superior performance to the A9, then under development (qv), could be produced by adopting the Christie type suspension and a powerful lightweight engine like the Liberty used in Christie's prototypes.

Funds were granted to buy two Christie vehicles, the first arriving from the United States, accompanied by Walter Christie, in the following month. Morris Commercial Cars Ltd acted as agents and licensees for the transaction and the vehicle was delivered as a "tractor" without a turret. The basic Christie chassis design incorporated compression spring suspension and large-diameter road wheels which could run either with or without the tracks. Trials led to the decision that, as far as the British were concerned, the "trackless" running facility could be done away with as an unnecessary complication. Also the Christie hull was too short and too narrow to take any existing (or contemplated) British turret. The power-to-weight ratio of the Christie design was $2\frac{1}{2}$ times better than the best existing British design, however, and it was decided to utilise the suspension but build a new chassis $5\frac{1}{2}$in wider and 10in longer, to take a 2pdr gun and turret. At the end of 1936 funds were allocated to build two prototypes, and Morris Commercial Cars Ltd were asked to undertake detailed design. The original Christie vehicle was now designated A13E1 and the two British-developed prototypes became A13E2 and A13E3.

The Liberty engine (an American World War I aero type) was adopted as standard, as in the original Christie vehicle, and Nuffields, an associate company of Morris, were to build it under licence. A13E2 was ready for trials in October 1937 and there followed a period of tests in which many mechanical problems were revealed, mostly due to the vehicle's high speed of over 35mph. Modifications included governing the speed down to 30mph, altering the clutch and transmission, and using shorter pitched tracks. By January 1938 most of the problems had been overcome and a production order (provisionally set at 50) was confirmed for 65 vehicles. Trials with A13E2, now joined by A13E3, were continued and further detail modifications were made to fittings before production was started by Nuffield Mechanisations and Aero Ltd, a company formed specially for munitions work by Morris. Deliveries started early in 1939, and the order was completed by Summer 1939. No further orders for this type were placed, since progress was being made with developments of the A13 design. This vehicle had taken only just over two years to get from inception to production status, a remarkably swift development for the period. These Cruisers Mk III as they were known were used by 1st Armoured Division in France in 1940, and (in small numbers) by 7th Armoured Division in Libya in 1940-41.

SPECIFICATION

Designation: Tank, Cruiser, Mk III (A13)
Crew: 4 (commander, gunner, loader, driver)
Battle weight: 31,360lb
Dimensions: Length: 19ft 9in; Height 8ft 6in; Width 8ft 4in; Track width 10⅛in; Track centres/tread 6ft 11in
Armament: Main: 1 × 2pdr QFSA
Secondary: 1 × Vickers ·303 cal MG
Armour thickness: Maximum 14mm
Minimum 6mm
Traverse: 360°. Elevation limits: —
Engine: Nuffield Liberty V12 340hp
Maximum speed: 30 mph
Maximum cross-country speed: 24mph (approx)
Suspension type: Christie
Road radius: 90 miles
Fording depth: 3ft
Vertical obstacle:
Trench crossing: 7ft 6in
Ammunition stowage: 87 rounds 2pdr.
3,750 rounds ·303 cal MG.
Special features/remarks: Turret similar to A9. Engine had dual starting system—electric or air compression.

CRUISER TANK, Mk IV and IVA (A13 Mk II) — United Kingdom

40. Cruiser Mk IV of 1st Armoured Division 1940; vehicle with additional armour plating over mantlet.

THE Cruiser Mk IV was essentially an uparmoured version of the Mk III and did, in fact, have the ordnance designation A13 Mk II. It followed the Cruiser Mk III in production and arose from a decision taken in early 1939 to increase the armour basis to 30 mm for cruiser tanks (for full circumstances see next entry, Covenanter). One of the A13 pilot models was accordingly reworked with additional armour to bring its thickness up to 20-30mm. Due to the high power-to-weight ratio of the basic design, there was little adverse effect on performance even though the weight was increased by more than 1,200lb. The extra armour plating was mainly on the nose, glacis, and turret front, but another feature was the addition of V-section armour plating on the turret sides which gave the "spaced armour" effect later widely used on German tanks. This resulted in the characteristic faceted turret sides, the feature by which the Mk IV could be most easily distinguished from the Mk III.

Nuffield undertook main production of the Cruiser Mk IV after Mk III production had been completed, starting in 1938. Some Mk IIIs were reworked with extra armour up to Mk IV standard and were externally similar, distinguished only by the early type mantlet as fitted to the Mk III. Mk IVA was the designation given to later production vehicles which had the Vickers co-axial machine gun replaced by a Besa. There was also a Mk IV CS which had a 3·7in mortar in place of the 2pdr gun. Only a small proportion of vehicles were of this type. Some vehicles had an armoured cover over the mantlet, and others were reworked with an armoured extension (heading picture) which completely concealed the mantlet. Cruisers Mk IV were used in France by 1st Armoured Division, 1940, and in the Western Desert by 7th Armoured Division, 1940-41. They were also used for training in Britain.

Total production of Cruiser Tank Mk IV series vehicles amounted to 655. Additional orders in 1939-40 were placed with the LMS, Leyland, and English Electric (200).

CRUISER Mk IV and IVA

41. Cruiser Mk III reworked to Mk IV standard.

42. Cruiser Mk IVA with armour cover on mantlet.

SPECIFICATION

Designation: Tank, Cruiser, Mk IV or IVA (A13 Mk II)
Crew: 4 (commander, gunner, loader, driver)
Battle weight: 33,040lb
Dimensions: Length 19ft 9in; Height 8ft 6in; Width 8ft 4in
Track width 10$\frac{3}{8}$in
Track centres/tread 6ft 11in
Armament: Main: 1 × 2pdr OQF
(1 × 3·7in howitzer in Mk IVCS)
Secondary: 1 × Vickers MG (Mk IV)
1 × Besa MG (Mk IVA)
Armour thickness: Maximum 30mm
Minimum 6mm
Traverse: 360°. Elevation limits:—
Engine: Nuffield Liberty V12 340hp
Maximum speed: 30mph
Maximum cross-country speed: 14mph (approx)
Suspension type: Christie
Road radius: 90 miles
Fording depth: 3ft
Vertical obstacle:
Trench crossing: 7ft 6in
Ammunition stowage: 87 rounds 2pdr
3,750 rounds ·303 cal or 7·92 cal MG
Special features/remarks: Spaced armour on turret sides.

43. Cruiser Mk IVA

44. Cruiser Mk V pilot model.

FOLLOWING his 1936 trip to Russia (which resulted in development of the original Christie-based A13) Lt Col Martel, Assistant Director of Mechanisation, suggested that in addition to a new fast cruiser tank there was need for a "medium" tank with 30mm armour and a high speed able to operate independently. This was obviously inspired by Martel's sight of the T-28 tank which the Russians had in large scale service, a design influenced by the British "16 tonner" of 1929, which had been abandoned on the grounds of expense. Specifications were drawn up and scale models made, with the final result that it was decided to build two pilot models with three-man turrets to a "simplified General Staff specification". These were designated A14 and A15 (later A16) respectively. The London Midland and Scottish Railway Company (LMS) built the former to plans worked out by the Chief Superintendent of Tank Design. This vehicle had Horstmann suspension and side skirt armour, a Thornycroft V12 engine and the newly designed Wilson compound epicyclic gearbox. The A16 was entrusted to Nuffields who had impressed Martel with the speed at which they had developed the A13. A16 was, in fact, virtually a heavier version of the A13. The A14 and A16 had superstructure layout and turret similar to the A9/A10 vehicles.

Meanwhile, the A9 had been uparmoured to a 30mm basis (becoming the A10) as a stop-gap, while the A14 and A16 were produced to the "medium" (or "heavy cruiser") requirements. Trials of the first A14 in early 1939 showed it to be noisy and mechanically complicated, as well as slower than the A13 pilot model fitted with an equivalent thickness of armour (see Cruiser Mk IV [A13 Mk II] entry). The LMS were therefore asked to slow up work on the A14 and work instead on the improved version of the A13—the A13 Mk III. This was to use as many A13 parts as possible, but it was to be built from the start with a 30mm armour basis, and have a lower overall height. A wooden mock-up was approved at the end of April 1939.

To keep the profile to the required low height, a Meadows Flat-12 engine (an enlarged version of that used in the Tetrarch light tank) was fitted, and the Wilson compound epicyclic gearbox (as used in the A14) was incorporated. Compared with the A13 Mk II—the Cruiser Mk IV—the driver's position was relocated to the right and the radiators for the engine were sited in the front of the vehicle on his left. The first production models were delivered in early 1940, but this vehicle proved a disappointment, due mainly to problems with the cooling system, which led to frequent breakdowns caused by the engine over-heating. Numerous modifications were necessary (see variants below) but the problem was never satisfactorily overcome. A less serious problem was high ground pressure due to the increased weight of this vehicle. In mid-1940 the vehicle was officially named Covenanter, the practice of naming British tanks dating from this time. Total output of Covenanters was 1,771, but the type was never used operationally, though it equipped UK-based armoured divisions for training until 1943. Some were sent to the Middle East for the training role, and others were converted to bridgelayers.

Work on the A14 and A16 was eventually abandoned in late 1939 before the second pilot model of each had been completed.

VARIANTS

Covenanter I (Cruiser Mk V): Basic production model produced by LMS.

Covenanter II (Cruiser Mk V*): Basic production model with service modifications to the engine cooling to overcome over-heating problems. Some vehicles had the armoured radiator louvres completely removed.

Covenanter III (Cruiser Mk V):** Vehicle with built-in engine cooling modifications, including vertical air louvres

COVENANTER

45. Covenanter ICS with 3 in howitzer. Note Bren AA mount on turret.

47. Covenanter Bridgelayer about to launch bridge.

at extreme rear. Stowage boxes added on nose each side of headlamp housing.

(Original designations given in parentheses.)

Covenanter IV: As Covenanter III but with additional built-in cooling modifications including air intake louvres on rear decking.

Covenanter CS (any mark): Small proportion of vehicles with 3in howitzer in place of 2pdr for close support work.

Covenanter Bridgelayer: Development of a 30ft 30 (long) ton folding ("scissors") bridge began in 1936 for mounting on a tank. Due to the availability and power of the Covenanter, a number of Mks I and II were fitted with the production type of scissors bridge, which was laid by hydraulic ram and arm installed in the fighting compartment, with power taken from the engine fan drive. Mainly used for training and development work, together with the Valentine Bridgelayer. The bridge was 34ft long overall and 9½ft wide. A few of these vehicles were used by the Australians in Burma in 1942.

Covenanter AMRA: A Covenanter was used in 1942 for tests with the newly developed Anti-Mine Roller Attachment (AMRA) which was intended to be pushed in front of a tank for mine-clearing. This was a trials installation only.

46. Covenanter IV showing typical appearance of production vehicles. Note mounts for stowage boxes on nose, and the cooling louvres. Cast mantlet cover was standard on production vehicles.

48. Covenanter AMRA on test.

Covenanter OP, Command, and ARV: A few vehicles were converted for these roles. Covenanter ARV had similar appearance to Crusader ARV (qv).

SPECIFICATION

Designation: Tank, Cruiser, Mk V, Covenanter (A13 Mk III)
Crew: 4 (commander, gunner, loader, driver)
Battle weight: 40,320lb
Dimensions: Length 19ft ⅜in — Track width 10¾in
Height 7ft 3¾in — Track centres/tread 7ft 6⅞in
Width 8ft 6¾in
Armament: Main: 1 × 2pdr OQF
(1 × 3in howitzer on CS models)
Secondary: 1 × Besa MG
Armour thickness: Maximum 40mm
Minimum 7mm
Traverse: 360°. Elevation limits: +20° to −15°
Engine: Meadows Flat-12 D.A.V. 280hp
Maximum speed: 31mph
Maximum cross-country speed: 25mph (approx)
Suspension type: Christie
Road radius: 100 miles
Fording depth: 3ft 2in
Vertical obstacle: 2ft 6in
Trench crossing: 7ft
Ammunition stowage: —
Special features/remarks: Similar external appearance to Cruiser Mk IV but lower overall height and with respaced road wheels. Riveted construction with cast mantlet cover. Fitted with Wilson compound epicyclic gears. 2in bomb thrower in turret.

49. Standard production Crusader I.

THE Crusader stemmed from the same line of development that gave rise to the Covenanter. The original A15 design for a "1938 Class Medium" was delayed in the planning stage due to uncertainties of requirements, and was re-designated A16 when Nuffields took on the project. Soon after approval of the A13 Mk III (Covenanter) mock-up was given in April 1939, the Director of Mechanisation asked the General Staff to review the alternative designs then being considered for the standard "heavy cruiser" role. These included the A18 (a larger development of the Tetrarch), the A14 (being developed by the LMS), the A16 (being developed by Nuffields), and the "new" A15, which was a proposed enlarged development of the A13 Mk III. The A15 emerged as the favoured design because (a) it utilised a large number of components from the A13 series, including the Christie suspension, (b) it could therefore be got into production faster, (c) it offered a better trench crossing ability than the A13 Mk III due to its increased length, (d) it was to be built to a 40-30mm armour basis, which was much superior to that of the other contenders. Nuffields were asked to produce a detailed design, based on the A13 Mk III but lengthened by one bogie wheel each side. In June 1939, Nuffields suggested that they use the Liberty engine of the original A13 design instead of the Meadows engine of the A13 Mk III since they already had the Liberty engine in production which would obviate delays. Since this also kept the weight down, the Director of Mechanisation agreed, and approval to go ahead was given in July 1939 with an initial order for 200 tanks plus the pilot model. This latter was ready by March 1940. In mid 1940 the order for A15s was increased to 400, then to 1062, and Nuffields became the "parent" company to a group of nine companies engaged in A15 production. Total output until 1943 was 5,300 vehicles.

"Teething" troubles with the pilot model included poor ventilation, inadequate engine cooling, and mechanical problems with the gear change.

Though most of the initial defects were overcome to an extent, the Crusader, as it was named in late 1940, always suffered from unreliability and the speed and urgency with which it was rushed into production did not allow long development trials, particularly for desert operations, where the Crusader became the principal British tank from Spring 1941 onwards. It first saw action near Capuzzo in June 1941, was prominent in all the major North African desert actions which followed, and was still in service, in its later 6pdr-armed form at the time of the Battle of Alamein in October 1942, though by then in the process of being displaced by American-built M3 and M4 mediums (qv). The last Crusaders in North Africa were finally withdrawn from first line use in May 1943, but the type was used for training until the end of the war. From mid 1942 onwards Crusaders were converted for numerous special purpose roles, including AA tanks, gun tractors, and ARVs. These are all detailed below.

The Crusader was designed just too late to incorporate any of the lessons learned in the early tank actions in France in 1940, but several modifications resulted from trials with the prototype. These included removal of the front auxiliary machine gun turret, mainly because it was too poorly ventilated and of limited value, which also simplified production. This turret was also removed retrospectively from many Mk I vehicles in service, and the space allowed extra ammunition stowage. It was also possible to increase the armour thickness slightly on hull and turret front. Finally, the Mk III version was up-gunned with a 6pdr replacing the 2pdr.

The Germans respected the Crusader for its speed, but it was no match for the PzKw III with 50mm gun, its main desert opponent, in hitting power, armour thickness, or serviceability. The German 55mm, 75mm and 88mm anti-tank guns also had no trouble in picking off Crusaders in the desert fighting.

VARIANTS

Crusader I (Cruiser Mk VI): Original production model with 2pdr gun and auxiliary front machine gun turret, which was later removed on some vehicles in service.

Crusader I CS (Cruiser Mk VI CS): As above but with 3in howitzer replacing 2pdr gun for close support role.

Crusader II (Cruiser Mk VIA): As Crusader I but with front machine gun turret eliminated during course of production programme. Extra frontal armour on turret and hull.

Crusader II CS (Cruiser Mk VIA CS): As Cruiser I CS with improvements as for standard Mk II.
(Original designations in parentheses.)

Crusader III: Final production version with 6pdr gun replacing 2pdr weapon, and increased armour on hull and turret and other parts of hull. Prototype tested November-December 1941. In production from May 1942 and 144 completed by July 1942.

Crusader OP, Crusader Command: Vehicles modified with dummy gun and extra radio and communications equipment for artillery and senior officers' use after Crusaders had been withdrawn from first line service.

Crusader III, AA Mk I: Crusader III with turret removed and replaced by single Bofors 40mm AA mount. In early conversions the unmodified ground mount was used, but most had an all-round open-topped shield.

Crusader III, AA Mk II: Crusader III with gun turret removed and replaced by new enclosed turret with twin 20mm Oerlikon AA cannon.

Crusader III, AA Mk III: As AA Mk II but with radio equipment removed from turret and installed in hull front next to driver.

Crusader AA with Triple Oerlikon: A few vehicles were converted to carry a triple 20mm Oerlikon AA cannon on an open mount. These appear to have been used for training only.

These AA versions were produced for the invasion of NW Europe in 1944, a troop of AA tanks being allotted to every HQ squadron. However, due to the air superiority of the Allied forces, and the consequent rarity of attacks by enemy aircraft, the AA troops were disbanded—shortly after the Normandy landings in June 1944.

Crusader II, Gun Tractor Mk I: Crusader II chassis with open-topped box superstructure and ammunition lockers, converted as a fast tractor for 17pdr anti-tank gun and its crew. Widely used by anti-tank regiments with armoured divisions in the NW Europe campaign, 1944-45. Side extensions could be fitted for deep wading in vehicles used by assault divisions in Operation Overlord.

Crusader ARV: Standard vehicle converted by removal of turret and addition of fittings for recovery equipment. Carried a demountable A-frame jib, and winch in former turret space.

50. Crusader IICS in the Western Desert, 1942.

Crusader Dozer: Conversion of standard tank for Royal Engineers use. Turret removed and winch and jib fitted for working dozer blade which was attached by frame on side of vehicle.

Crusader Dozer and Crane (ROF): Adaptation of the standard Crusader Dozer for Royal Ordnance Factory use, handling unexploded shells and bombs. Dozer fixed in raised position as support for additional armour, and extra armour protection added on hull top.

Crusader with AMRA Mk Id: Crusader tank fitted to push Anti-Mine Roller Attachment (AMRA). Appearance similar to Covenanter with AMRA (qv). Not used operationally.

Crusaders were also used for various experimental work. One was tested with a form of flotation equipment based on the use of catamaran attachments. A Crusader chassis was also fitted experimentally post-war with a 5·5in gun as a test mount for a similar installation planned for the Centurion.

SPECIFICATION

Designation: Tank, Cruiser, Mk VI, Crusader I or II
Tank, Cruiser, Crusader III
Crew: 3–5 (commander, gunner, driver (Mk III) plus loader and hull gunner (Mks I and II)
Battle weight: 42,560lb (Mks I and II)
44,240lb (Mk III)
Dimensions: Length 19ft 8in; Height 7ft 4in; Width 8ft 8in; Track width 10¾in; Track centres/tread 7ft 7in
Armament: Main: 1 × 2pdr OQF (Mks I and II)
1 × 6pdr OQF (Mk III)
Secondary: 1 or 2 × Besa 7·92 cal MG
Armour thickness: Maximum 40, 49, 51mm (Mks I, II, III in order)
Minimum 7mm
Traverse: 360°. Elevation limits: —
Engine: Nuffield Liberty V12 340hp
Maximum speed: 27mph
Maximum cross-country speed: 15mph (approx)
Suspension type: Christie
Road radius: 100 miles (127 miles with extra fuel tank at rear)
Fording depth: 3ft 3in
Vertical obstacle: 2ft 3in
Trench crossing: 8ft 6in
Ammunition stowage: 110 rounds, 2pdr; 65 rounds 6pdr (Mk III)
5,000 rounds, MG (max)
Special features/remarks: Riveted hull and welded turret of composite construction with outer layer of armour bolted on. Many Crusaders in the Western Desert were fitted with prominent "skirted" track covers. Data refers to basic marks with 2pdr and 6pdr guns.

51. Crusader III. Note twin Vickers "K" gun AA mount on turret.

52. Crusader II, AA Mk I. Later pattern with full shield to gun.

53. Crusader III, AA Mk II. AA Mk III of similar appearance.

54. Crusader AA with triple Oerlikon mount.

55. Crusader ARV with A-frame jib shown erected and twin Bren AA mount.

56. Crusader Dozer.

57. Crusader II, Gun Tractor Mk I.

58. Crusader Dozer and Crane, Royal Ordnance Factory.

59. One of the six Cavalier pilot vehicles.

EXPERIENCE with the Covenanter and the Crusader pilot model in late 1940, plus the accrued lessons of the tank actions in France and Libya led the Ministry of Supply, at the end of 1940, to ask for a heavy cruiser tank which overcame the inherent faults of the existing designs. Thicker armour (65mm on hull front, 75mm on turret front), bigger turret ring (60in diameter), a heavier gun (6pdr), a more powerful engine, a weight limit of 24 tons, a speed of not less than 24mph, and above all much better mechanical reliability, were among the many improvements requested. In January 1941 the Tank Board considered proposals to meet these requirements. Vauxhall offered a scaled down version of the Churchill infantry tank (qv), designated A23, which was not taken up, while Nuffield Mechanisations and Aero offered a design based upon that of the Crusader which they were already building. Mechanisation and Aero were therefore asked to build six pilot vehicles for completion by the following autumn, under the designation A24. This vehicle was mechanically similar to the Crusader and was to retain the Liberty engine and Wilson epicyclic gearbox.

Meanwhile, Leyland Motors, one of the production group involved in building the Covenanter and Crusader, suggested a design which utilised a chassis similar to the Crusader but was powered with an adapted version of the Rolls-Royce Merlin aero engine and had the new Merritt-Brown gearbox already being used in the Churchill. However, at this period all Merlin production was needed for aircraft, but the basic chassis design of the A24 was rationalised with the Leyland design (then designated A27) and an order for 500 A24s was placed "off the drawing board" in June 1941, and the name Cromwell I was allocated. The A24 was to be considered as an interim type while the Merlin-engined A27 was developed. It was then decided to reserve the name Cromwell for the A27 and the name Cavalier was given to the A24, first pilot model of which was eventually completed in January 1942.

With its Liberty V12 engine, the A24 Cavalier offered no mechanical advance over the Crusader and, in fact, due to the increased weight it had an inferior performance, shorter engine life, and was even more prone to breakdown. Production vehicles were used only for training as gun tanks, but in 1943 half were converted to OP tanks for artillery use and in this guise some were used by artillery regiments of armoured divisions in the NW Europe campaign of 1944-45. A few others were converted to ARVs.

Variants

Cavalier OP: 6pdr gun replaced by dummy barrel. Appearance as Centaur OP and Cromwell OP (qv).

Cavalier ARV: Turret removed and winch fitted in turret space. Carried demountable A-frame jib. Appearance as Centaur and Cromwell ARV (qv).

Specification

Designation: Tank, Cruiser, Mk VII, Cavalier (A24)
Crew: 5 (commander, gunner, loader, driver, co-driver)
Battle weight: 59,360lb
Dimensions: Length 20ft 10in; Height 8ft; Width 9ft 5½in; Track width 14in; Track centres/tread 8ft 1¾in
Armament: Main: 1 × 6pdr OQF
Secondary: 1 or 2 Besa 7·92 cal MG
Armour thickness: Maximum 76mm
Minimum 20 mm
Traverse: 360° Elevation limits: —
Engine: Nuffield V12 410hp
Maximum speed: 24mph
Maximum cross-country speed: 14mph (approx)
Suspension type: Improved Christie
Road radius: 165 miles
Fording depth: 3ft
Vertical obstacle: 3ft
Trench crossing: 7ft 6in
Ammunition stowage: 64 rounds 6pdr
4,950 rounds 7·92 cal MG
Special features/ remarks: Externally almost identical to Cromwell and Centaur. Suspension improved with stronger coil springs to compensate for extra weight compared to Crusader. Uprated Liberty engine. Some vehicles lacked hull machine gun. Fitted with Mk 3 or Mk 5 6pdr, the latter distinguished by prominent counter-weight on muzzle—and lighter appearance.

60. Early production Centaur with Mk 5 6pdr gun and hull machine gun not fitted.

ESSENTIAL feature of Leyland's A27 design was the Rolls-Royce Merlin aero engine, which in its form adapted for tank use was named the Meteor. With an output of 600hp, this power unit gave ample performance for the heavy cruiser category of tank, was a far superior engine to the Liberty (which dated basically to 1917), and was already in production for aircraft use which meant that valuable time was saved which would otherwise have been necessary for development work. Birmingham Carriage & Wagon were asked to become "parents" for the A27 and work out design details. However, due to the shortage of Merlin engines, needed for aircraft production, an interim version of the A27 was proposed, retaining the Liberty engine of the Crusader, but otherwise identical to the Meteor-engined design. English Electric were asked in November 1941 to design this version, which was designated A27L (L: Liberty engine), Cromwell II. The Meteor-engined variant was to be called Cromwell III. Leyland took over production "parentage" and the name was meanwhile changed to Centaur, the Cromwell name being kept exclusively for the Meteor-engined A27s. The A27L design had to be to the same dimensions as the Nuffield-designed Cavalier, but the Liberty engine installation had to be interchangeable with the Meteor power plant to allow for re-engining at a later date when Meteors became available in quantity. A Merritt-Brown gearbox was used instead of the Wilson epicyclic gearbox (to conform with the Meteor-engined design), and other changes, compared with the Cavalier, involved shifting the radiators to the rear of the engine compartment and incorporating mechanical parts common to the Meteor-engined vehicle. The Centaur pilot model was completed in June 1942 and first production vehicles were ready by the end of the same year. About 950 were completed, 80 of them as close support models with a 95mm howitzer in place of the 6pdr gun. These were used on the D-Day landings by the Royal Marines Armoured Support Group, giving covering fire from LCTs and then landing over the beaches, on June 6, 1944.

Remaining vehicles were either converted to Cromwells by the installation of the Meteor engine in 1943, becoming the Cromwell X, (later Cromwell III), or were used for training in their original form. Others were converted for special purpose roles, including AA tanks, ARVs, OP vehicles, and Dozers, all used in the NW Europe campaign. These are detailed below.

VARIANTS

Centaur I: Original production version with 6pdr gun.

Centaur III: Version with 75mm gun (equivalent to Cromwell IV). Some were converted for special purpose roles—see below—and others converted to Cromwells.

Centaur IV: Close support model with 95mm howitzer. 80 built. Up-rated engine.

Centaur OP: Conversion with dummy gun for artillery observation officers' use. Extra radio equipment fitted, plus telephones, etc.

Centaur, AA Mk I: New turret as Crusader AA Mk II (qv) but with Polsten instead of Oerlikon cannon.

Centaur, AA Mk II: Turret as Crusader AA Mk III (qv) but with Polsten instead of Oerlikon cannon.

Both these versions were based on Centaur III or IV chassis.

Centaur ARV: Turret removed and winch fitted in turret space. Fitted to carry demountable A-frame jib.

Centaur Kangaroo: Centaur with turret removed used as armoured personnel carrier. Few only so converted.

Centaur Dozer: Vehicle with turret removed, and winch and jib fitted to work dozer blade attached on side frames. Allocated one per squadron to regiments equipped with Cromwells and also used by Royal Engineers.

61. Centaur IV (95mm howitzer) of Royal Marines at Normandy, June 1944.

SPECIFICATION
Designation: Tank, Cruiser, Mk VIII, Centaur (A27L)
Crew: 5 (commander, gunner, loader, driver, co-driver)
Battle weight: 63,600lb
Dimensions: Length 20ft 10in   Track width 14in
Height 8ft 2in   Track centres/tread 8ft $1\frac{3}{4}$in
Width 9ft 6in
Armament: Main: 1 × 6pdr OQF (Mk I)
1 × 95mm howitzer (Mk IV)
Secondary: 1 or 2 × Besa 7·92 cal MG
Armour thickness: Maximum 76mm
Minimum 20mm
Traverse: 360°. Elevation limits: —
Engine: Liberty V12 395hp
Maximum speed: 27mph
Maximum cross-country speed: 16mph (approx)
Suspension type: Improved Christie
Road radius: 165 miles
Fording depth: 3ft
Vertical obstacle: 3ft
Trench crossing: 7ft 6in
Ammunition stowage: 51 rounds 95mm (Mk IV)
64 rounds 6pdr (Mk I).
4,950 rounds 7·92 cal MG
Special features/remarks: Externally similar to Cavalier and Cromwell. Uprated Liberty engine with new version (improved design built by Leyland) in Mks III and IV.

62. Centaur OP in use as a Brigade Commander's tank, Germany, April 1945. Note stays on dummy barrel.

63. Centaur AA Mk II. Note removal of hull machine gun. Radio was located in hull gunner's position.

64. Centaur Dozer with blade in lowered position.

65. Centaur ARV with A-frame jib erected.

66. Rear-view of Centaur AA Mk II showing the flush-topped rear decking which distinguished the A27L from the A27M. The Cromwell (A27M) had a raised armoured louvre behind the turret on the rear decking.

67. Cromwell VIIw. Note appliqué armour and 75mm gun.

INITIALLY known as Cromwell III, the Meteor-engined version of the A27 design was designated A27M (M: Meteor engine). The Meteor engine as adapted from the Merlin for tank use had about 80% of its component parts identical to the aircraft engine, thus greatly facilitating production for tanks. Rolls-Royce converted a batch of Merlins for use in tanks, and during 1941 two Crusader tanks had Meteor engines installed in place of their Liberty power plants for exhaustive test running. This enabled positioning of auxiliary components, wear and tear, oil consumption, and so on to be determined at an early date while design work on the A27 itself proceeded. Birmingham Carriage & Wagon delivered the first mild steel prototype to the Army for trials on March 1, 1942, actually ahead of the Centaur pilot model. Two more pilot models were delivered by the end of 1942, and teething troubles on tests proved relatively minor—mainly detail points concerned with clutch, gears, and cooling. The idea of using the powerful Meteor engine was handsomely vindicated by results, and ample power was available for any forseen developments of the A27 type. Cromwell production started in January 1943, by which time Leyland had become the design and production "parents" for the entire A27 series. This embraced all subcontractors for component parts as well as plants building Cromwells.

Meanwhile, War Office policy with regards to tank armament had changed considerably since the "heavy cruiser" requirement resulting in the A24/A27 series had been formulated. Fighting in the Western Desert, coupled by the decisive appearance of the American-built M3 and M4 Medium tanks in that theatre led to a requirement for a gun with "dual-purpose" capability—able to fire HE or AP shot—as fitted in the very successful M3 and M4 mediums. Work on a British designed version of the 75mm gun virtually a bored-out development of the British 6pdr able to fire American ammunition, was put in hand in December 1942 and Cromwells from Mk IV onward were produced with this weapon in place of the 6pdr. The first vehicles so equipped were delivered in November 1943, but there were many initial defects in this gun, including unsatisfactory semi-automatic cams in the breech, which were not entirely put right until May 1944.

The Cromwell was numerically the most important British-built cruiser tank of World War II, forming the main equipment of British armoured divisions in 1944-45 together with the American-built M4 Sherman. However, even with a 75mm gun it was still, by 1944 standards, inferior to contemporary German tanks like the Panther and late-model PzKw IVs. With its Meteor engine it was the fastest and most powerful of British tank designs until that period, but physical limitations (mainly the narrowness of the hull) prevented its being upgunned further and considerable redesign was necessary to turn it into a vehicle capable of carrying the very desirable 17pdr gun armament—see the Challenger and Comet for further details.

All the A24/A27 series were structurally similar, with a hull and turret of simple box shape and composite construction—an inner skin with an outer layer of armour bolted on. Driver and co-driver/hull machine gunner sat in the forward compartment, and the turret crew consisted of the commander, gunner and loader who was also the radio operator. Tracks were manganese with centre guides, and the engine and transmission were at the rear. Numerous detail modifications were incorporated during the Cromwell's production run, which ended in 1945. These are noted below. Most important innovation was the introduction of all-welded construction in place of rivetting on later models, thus further simplifying mass-production.

CROMWELL

VARIANTS

Cromwell I: Original production model with 6pdr gun. Similar in external appearance to Centaur I.

Cromwell II: Mk I modified by removal of hull machine gun and fitting of wider tracks—$15\frac{1}{2}$in in place of 14in.

Cromwell III: Centaur I re-engined with Meteor to bring it to A27M standards. Originally designated Cromwell X.

Cromwell IV: Centaur III with 75mm gun re-engined with Meteor.

Cromwell IVw: As Mk IV but with all-welded hull, and built with Meteor engine.

Cromwell Vw: As Mk IV but with all-welded hull.

Cromwell VI: As Mk IV but with 95mm howitzer replacing the 6pdr for the close support role.

Cromwell VII: Cromwell IV re-worked with appliqué armour welded on hull front, $15\frac{1}{2}$in tracks replacing 14in tracks, stronger suspension, and reduced final drive ratio to govern down maximum speed to 32mph.

Cromwell VIIw: Cromwell Vw modified as above.

Cromwell VIII: Cromwell VI modified as above.

Cromwell ARV: Vehicle with turret removed, winch fitted in turret space, and demountable A-frame jib. Appearance as Centaur ARV (qv).

Cromwell Command/OP: Mk IV, VI or VIII fitted with dummy gun and extra radio equipment for use of formation commander or artillery observation officers.

Cromwell CIRD: Vehicle with fittings to take Canadian Indestructible Roller Device (CIRD) mine exploding equipment. Few only converted.

Cromwell "Prong": Standard vehicle fitted with Cullin Hedgerow Cutting Device, Normandy 1944. This equipment, at first, extemporised in the field, then manufactured in limited quantities, was fitted to some Cromwell and Sherman tanks in June-August 1944 to assist in breaking through the extensive hedges and foliage of the "Bocage" country of Normandy which otherwise tended to restrict movement to the roads.

Charioteer: Post-war (1950) conversion of existing Cromwell chassis with new turret and 20pdr gun.

Several experimental or trials models of Cromwells were produced to test installations or proposed modifications. Three of these are illustrated. Also projected was a Cromwell Crocodile, still under development at the war's end. It was similar to the Churchill Crocodile (qv).

SPECIFICATION

Designation: Tank, Cruiser, Mk VIII, Cromwell (A27M)
Crew: 5 (commander, gunner, loader, driver, co-driver)
Battle weight: 61,600lb
Dimensions: Length 20ft 10in; Height 8ft 2in; Width 9ft $6\frac{1}{2}$in (10ft, Mks VII, VIII); Track width 14in ($15\frac{1}{2}$in later marks); Track centres/tread 8ft $1\frac{3}{4}$in
Armament: Main: 1 × 95mm howitzer (Mks VI, VIII)
1 × 6pdr (Mks I–III)
1 × 75mm (Mks IV, V, VII)
Secondary: 2 × Besa 7·92 cal MG (1 on Mk II)
Armour thickness: Maximum 76mm (101mm with appliqué armour)
Minimum 8mm (10mm on welded variants)
Traverse: 360°. Elevation limits: +20° to $-12\frac{1}{2}$°
Engine: Rolls-Royce Meteor V12 600hp
Maximum speed: 40mph (32mph from late Mk IVs on)
Maximum cross-country speed: 18mph (approx)
Suspension type: Improved Christie
Road radius: 173 miles
Fording depth: 3ft (4ft prepared)
Vertical obstacle: 3ft
Trench crossing: 7ft 6in
Ammunition stowage: 64 rounds, main armament
4,950 rounds, cal MG
Special features/remarks: Frequently seen with flame cover over rear exhaust. Production variations: Type C—with Valentine type axle. Type D—escape hatch added in hull for co-driver. Type F—escape hatch added for driver. Type E—vehicle with final drive ratio reduced. Vehicles with appliqué armour had weight increased by approximately 1120lb. Late production vehicles had Comet type exhaust modifications.

68. Cromwell III (originally Mk X) was a Centaur re-engined with a Meteor unit.

69. Cromwell I; this particular vehicle is one of thirty pre-production Cromwells in mild steel for training and trials.

70. Cromwell IV. This is a late production vehicle with F type hull which had escape doors (seen open) for driver and co-driver.

71. Cromwell VI was the close support variant with 95mm howitzer.

72. Cromwell IV command tank used by Div HQ of 11th Armd Divn. followed by a Centaur OP with dummy gun.

73. Cromwell "Prong" was the name given to vehicles fitted with Cullin Hedgerow Cutting Device. This picture shows the test installation on a Centaur III with the 75mm gun.

74. Cromwell IV with CIRD. This vehicle has a D type hull with co-driver's escape door only in hull.

75. Cromwell II used as a test vehicle for trial installation of appliqué armour to hull front and turret.

76. Vauxhall-built Cromwell I pilot model with cast/welded turret similar to the type used on the Churchill. Vauxhall were to build Cromwells when Churchill production was scheduled to cease in 1943. However, it was decided to continue Churchill production and Vauxhall withdrew from the Cromwell programme.

77. Cromwell III, D type hull, with addition of co-driver's escape hatch, and fitted with side skirts for desert service. Though produced as a "standard" fitting, these skirts were rarely fitted. This view shows the trial installation.

78. Standard production Challenger.

BRITISH reverses in the great tank engagements against the Afrika Korps in the Western Desert in 1941 led the British General Staff to ask the Tank Board to investigate the feasibility of mounting a heavy high velocity gun (able to knock out any known German AFV) in British infantry and cruiser tanks. For the infantry tank requirement it was proposed to fit a 3in AA gun in a limited traverse mount in the Churchill chassis, though ultimately this idea was abandoned (see Churchill history). The 17pdr gun was then in the development stage, and for the cruiser tank requirement the possibility of mounting this weapon in the A27 series (qv) was considered. However, the A27 chassis was too narrow to take a turret big enough to hold the 17pdr, so Birmingham Carriage & Wagon, then the A27 design "parents", evolved a design based closely on the A27, but with a lengthened hull, an extra bogie wheel each side to compensate for the extra weight, and a widened centre hull section. Work on three pilot models was under way by May 1942, and the first was delivered for trials the following August. The A30, as the new vehicle was designated, was based as nearly as possible on Cromwell components. Mechanically it was almost identical, and it had the same Rolls-Royce Meteor engine. The turret to hold the 17pdr gun was designed and built by Stothert & Pitt and was derived very closely from the turret they had tested on TOG 2 (qv).

Trials with the prototype showed up several deficiencies; the extra weight, 3 (long) tons more than the Cromwell, led to suspension trouble; there was heavy trunnion loading and slow turret traverse. To overcome the latter defect, electric Metadyne type traverse gear was subsequently installed. The size of 17pdr AP rounds, plus the size of the turret and breech, severely restricted ammunition stowage, and the hull machine gun position was eliminated to provide more room. A further modification found necessary was the reduction of the armour thickness to reduce weight and so improve performance. In January 1943, the prototype trials were assessed to see if a production order was justified. By this time the PzKw IV with "long" 75mm gun had appeared with the Tiger soon to follow, and the fact that the A30 design was a quick way of getting a tank with 17pdr gun into service swung the balance in favour of its adoption. An order for 200 vehicles was placed, these to be distributed to armoured divisions in similar proportions to AA and close support tanks. This would mean that divisional commanders would have available a tank capable of taking on German vehicles with 75mm and 88mm guns on almost equal terms as far as fire power was concerned. In view of the adverse qualities revealed by the trials, however, it was decided to investigate the possibility of mounting the 17pdr in the M4 Sherman (by then in large scale British service) as a safeguard against failure of the A30 programme.

Numerous troubles were experienced in getting the A30, now named Challenger, into production. It was not until March 1944 that the first production vehicles were ready, by which time it was realised that no provision for waterproofing for deep wading existed. This was essential for getting ashore from LCTs for the forthcoming Overlord landings (June 6, 1944). Meanwhile the Sherman with the 17pdr—known as the Firefly—was available and was adopted in place of the Challenger. Later in 1944, however, some Challengers were issued to the reconnaissance regiments of British armoured divisions in NW Europe to stiffen up the fire power of the 75mm-armed Cromwells with which they were equipped.

SPECIFICATION

Designation: Tank, Cruiser, Challenger (A30).
Crew: 5 (commander, driver, gunner, loaders (2))
Battle weight: 72,800lb
Dimensions: Length 26ft 8¾in
Height 9ft 1¼in
Width 9ft 6½in
Track width 15½in
Track centres/tread 8ft 1¾in

79. Rear view of Challenger, turret traversed, clearly shows widened centre hull.

80. Challenger in service in France, 1944.

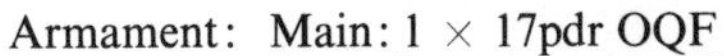

Armament: Main: 1 × 17pdr OQF
Secondary: 1 × ·30 cal Browning MG (co-axial)
Armour thickness: Maximum 101mm
Minimum 10mm
Traverse: 360°. Elevation limits: +20° to −10°
Engine: Rolls-Royce Meteor V12 600hp
Maximum speed: 32mph
Maximum cross-country speed: 15mph (approx)
Suspension type: Improved Christie
Road radius: 120 miles
Fording depth: 4ft 6in (prepared)
Vertical obstacle: 3ft
Trench crossing: 8ft 6in
Ammunition stowage: 42 rounds, 17pdr
Special features/remarks: Strong family resemblance to A27 series, but distinguished by high turret, long gun, extra hull length and extra bogie wheels. Sound project which was plagued by troubles due to necessity for virtual re-design of original A27 type hull and features to accommodate the wider turret. Production vehicles had slightly smaller bogie wheels than A27 series. Some vehicles had 25mm appliqué armour on turret front and rear, and on hull front.

81. Challenger in service with rear dustguards removed. Note door in turret rear.

82. Late production Challenger; note the splash plate protecting the turret ring. This was a retrospective modification added as a result of trials.

83. Standard production Avenger; note the open-topped turret.

DEVELOPMENT of the 17pdr anti-tank gun started in late 1941 and it was approved for production by mid 1942, by which time the first pilot model of the A30 Challenger was nearing completion as an adaptation on the A27 design to take this new weapon. The Germans were by this period making wide use of "tank destroyers"—large calibre guns in limited traverse, armoured self-propelled mounts—and the British General Staff thought it desirable that vehicles on similar lines, mounting the largest practicable anti-tank gun, should be produced to equip the anti-tank battalions of armoured divisions. It was hoped to acquire the American M10 GMC tank destroyer at first, but this vehicle was only just going into production for the US Army and no deliveries could be promised for Britain before the end of 1943.

Alternative designs were therefore investigated involving the mounting of a 17pdr in existing tank chassis. The Crusader chassis was clearly too small and underpowered, but the chassis of the Valentine tank (qv), the only readily available alternative, could be adapted. This involved a rear-facing mount for the 17pdr, which was not entirely satisfactory, but the Valentine SP, known as the Archer, could be got quickly into production as a "stop gap" design while something better was evolved.

The obvious choice was an adaptation of the A30, and early in 1943, while the Archer pilot model was still being built, Leylands, the "parent" company for the A27 series, were asked to produce an SP variant of the A30. The size of the A30 chassis allowed the desirable facility of all-round traverse as featured in the US M10 and in all mechanical respects the A30 SP was identical to the A30 tank. However, Leyland were concurrently working on the A34 (Comet) design as a development of the A27 series, and the return rollers featured in the A34 suspension were also incorporated in the A30 SP design. The prototype vehicle, however, had the same suspension as the Challenger. Named Avenger, the A30 SP also differed from the Challenger in having its superstructure height reduced to lower the overall height of the vehicle.

The turret had a mild steel canopy (incorporating hatches) for crew protection. By the time the pilot model A30 SP was ready in 1944, the M10 was coming into service with British "tank destroyer" battalions, and priority was being given to Comet tank production. An order for about 230 Avengers was not fulfilled until 1946 and the type briefly equipped two SP artillery battalions after the war.

SPECIFICATION

Designation: SP 17pdr, A30 (Avenger)
Crew: 4–5 (commander, driver, gunner, loader (optional second loader))
Battle weight: 69,440lb
Power/weight ratio: 22hp/ton
Dimensions: Length 28ft 7in (24ft 3in gun aft) Track width 15½in
Track centres/tread 8ft 1¾in
Height 7ft 3in
Width 10ft
Armament: Main: 1 × 17pdr OQF
Secondary: 1 × ·303 cal Bren MG (AA)
Armour thickness: Maximum 101mm
Minimum 10mm
Traverse: 360°. Elevation limits: +20° to −10°
Engine: Rolls-Royce Meteor V12 600hp
Maximum speed: 32·3mph
Maximum cross-country speed: 20mph (approx)
Suspension type: Improved Christie with return rollers.
Road radius: 105 miles
Fording depth: 3ft (4ft prepared)
Vertical obstacle: 3ft
Trench crossing: 8ft 6in
Ammunition stowage: 55 rounds, 17pdr
Special features/remarks: Distinguished from A30 Challenger by return rollers and canopy on open turret. Turret had vertical sides, welded, with cast front. Heavy casting on turret rear acted as counterweight to gun. Canvas screens below turret canopy gave weather protection. No vision devices for driver since vehicle was not intended to fire on the move. Turret had Metadyne or hand traverse as Challenger.

84. Avenger with driver's canopy and windscreen erected for road running.

85. Avenger from rear with turret counter-weight fully visible.

86. Avenger from front, fully closed down.

87. The Avenger prototype, showing the Challenger-type suspension fitted on this vehicle.

88. Standard production Comet.

THOUGH Birmingham Carriage & Wagon were the original design and production "parents" for the A27 (Centaur/Cromwell) series, production delays and difficulties in correcting mechanical deficiencies led to the parentage being transferred to Leylands in May 1943. The quest to upgun the Cromwell led to the development of the A30 Challenger with 17pdr gun, which, as had been noted, was not a satisfactory design.

When Leyland took over the A27, they immediately began work on designing an improved version incorporating modifications to overcome the limitations of the A27 design. Mechanically and dimensionally the improved vehicle, designated A34, was similar to the A27. To avoid the need for widening the hull, as had been necessary in the A30 to take the 17pdr gun, Vickers-Armstrong designed a new "compact" version of the 17pdr with a shorter barrel, shorter breech, and lighter weight. Known originally as the Vickers HV 75mm (HV: High Velocity), but later called the 77mm gun, it had a performance and penetrating power only slightly inferior to the 17pdr gun and fired the same ammunition. It was intended that the 77mm gun would fit the original A27 type turret with only small modifications, but in the event an enlarged turret was found necessary. The A34 pilot model was ready for tests in February 1944, and among modifications incorporated as a result of trials was a stronger suspension with the addition of track return rollers. By this time the A34 represented a 60% re-design of the A27 and was virtually a new tank. Production deliveries commenced in September 1944 and the first A34s, then named Comet, were issued to battalions of 11th Armoured Division after the Rhine crossing in March 1945, though a few vehicles were issued earlier, in December 1944.

The Comet proved a fast and reliable tank, the first British AFV to come near matching the German Panther in performance and gun power. However, it appeared too late to play any prominent part in British tank combat in World War II. Comets subsequently equipped British armoured units post-war and some remained in service until the early sixties. Like the later marks of Cromwell, the Comet was of all-welded construction.

SPECIFICATION

Designation: Tank, Cruiser, Comet I (A34)
Crew: 5 (commander, gunner, loader, driver, co-driver)
Battle weight: 78,800lb
Dimensions: Length 25ft 1½in (21ft 6in, gun aft) Track width: 15½in
Width 10ft
Height 8ft 9½in Track centres/tread 8ft 1¼in
Armament: Main: 1 × 77mm OQF
Secondary: 2 × Besa 7·92 cal MG (one co-axial)
Armour thickness: Maximum 101mm
Minimum 14mm
Traverse: 360°. Elevation limits: +20° to −12°
Engine: Rolls-Royce Meteor V12 600hp
Maximum speed: 29mph
Maximum cross-country speed: 16mph (approx)
Suspension type: Improved Christie with return rollers
Road radius: 123 miles
Fording depth: 3ft (4ft prepared)
Vertical obstacle: 3ft
Trench crossing: 7ft 6in
Ammunition stowage: 58 rounds 77mm
Special features/remarks: Distinguished from Cromwell by return rollers above road wheels. Welded turret with cast front, prominent cupola, and forward sloping roof. Prominent counterweight at turret rear. Late production vehicles had armoured fishtail exhausts and improved "breathing" system.

89. Mild steel pilot model of the A34 showing the original Cromwell-type suspension.

90. Side view of fully-stowed Comet shows clearly the all-round vision cupola and the front hull escape door.

91. Comet from rear showing hull and turret top details.

92. Centurion (A41) armed with 17pdr gun and Polsten cannon in ball-mount. This vehicle was later designated Centurion I.

EXPERIENCE in the Western Desert fighting of 1941-42 had a profound effect on subsequent British tank policy; in particular the desert fighting finally shattered the old tank versus tank (ie, "comparable Class") theory and showed that tanks were just as likely to have to support infantry or attack anti-tank guns with HE fire as to engage other tanks. The American M3 medium tank, which was issued to the British in the desert in 1942, effectively fulfilled the need for an AFV which could fire either AP or HE shot as required. Largely resulting from the desert war tank combats, the War Office revised its policy for future tank development in September 1942 and called for an "all-purpose" or "universal" tank chassis which could be developed to fulfil the various roles previously carried out by several unrelated chassis designs. This initiated the thinking which led to the Centurion tank. Meanwhile the old weight and dimension limitations to conform with British railroad loading gauge had been lifted by the War Office under pressure from the Department of Tank Design, and they were thus able to undertake initial design studies on this basis. At this period, however, the Government had banned development work on projects which could not be in service by 1944 (this was to concentrate work on perfecting existing designs) and authority to proceed was not given until July 1943. AEC were appointed production "parents" and the new vehicle, designated A41, was to be produced in the first instance for the "heavy cruiser" role. It was required to mount the largest calibre tank gun (the 17pdr), have a sloped instead of vertical glacis to improve frontal protection, and be sufficiently armoured to withstand the German 88mm gun. Road speed was less important than cross-country performance which had to match that of the Comet at least.

A mock-up of the design was ready by May 1944 and featured modified Horstmann suspension in place of the Christie type of previous cruiser types. This was because increasing weight had now exceeded the effectiveness of the Christie suspension. The hull gunner's position was omitted to increase ammunition stowage, and the hull was boat-shaped to improve resistance to mine explosions. The Meteor engine was again used in this vehicle, together with a Morris 8HP auxiliary motor to charge the dynamo and work the fans. Merrit-Brown gearbox was standard except in the A41S (see below).

Twenty pilot models were ordered with 17pdr guns (though the last five were to have 77mm guns) with various combinations of Polsten cannon and Besa machine gun as secondary armament. Pilot models 1-10 had Polstens and rear turret escape doors; pilots 11-15 had Besa MGs instead of Polstens; pilots 16-18 had an additional Besa MG in a ball mount in place of the rear turret escape door; and pilots 19-20 had provision for mounting a Besa MG in the hull front, reverting to the escape door in the turret rear. The last five vehicles had "Powerflow" gearboxes and were designated A41S. The first six vehicles produced, later called Centurion I, were delivered in May 1945 and rushed to Germany for testing in combat conditions with 22nd Armoured Brigade. However, hostilities had ceased by the time they arrived. Meanwhile in January 1945, an up-armoured vehicle, designated A41A, Centurion II, was produced as a prototype, also armed with a 17pdr gun. This vehicle had a new cast turret and numerous detail improvements. A further projected version was to have a 95mm howitzer for the close support role, though this was never built.

The Centurion subsequently became the standard British post-war battle tank and later variants and marks were still in widespread service in 1969. Development beyond 1945 is outside the scope of this book, however. The Centurion I never went into full production and the Centurion II (later Centurion 2) became the first production model.

SPECIFICATION

Designation: Tank, Cruiser, Centurion I (A41)
Crew: 4 (commander, driver, gunner, loader)
Battle weight: 107,520lb
Dimensions: Length 25ft 2in Track width 24in

93. A41 first pilot model.

94. A41 Centurion second pilot model, fitted with stowage boxes on turret.

Height 9ft 8in   Track centres/tread 8ft 8in
Width 11ft
Armament: Main: 1 × 17pdr OQF
Secondary: 1 × 20mm Polsten cannon or
1 × 7·92 cal Besa MG in turret front
Armour thickness: Maximum 152mm (Mk II)
Minimum 17mm
Traverse: 360°. Elevation limits: +45° to −15°
Engine: Rolls-Royce Meteor V12 600hp
Maximum speed: 21·4 mph
Minimum cross-country speed: 15mph (approx)
Suspension type: Modified Horstmann
Road radius: 60 miles
Fording depth: 4ft 9in
Vertical obstacle: 3ft
Trench crossing: 11ft
Ammunition stowage: 70 rounds 17pdr.
Special features/remarks: The Polsten cannon or Besa MG in turret front had optional linkage to main armament. First British designed tank with stabiliser for main armament. All-welded construction. Welded turret in Mk I and cast turret in Mk II.

95. A41 second pilot model from rear showing turret escape door.

96. A41, first pilot, from side.

INFANTRY TANK, Mk I, MATILDA I (A11)

United Kingdom

97. Standard production Matilda I.

IN April 1934, the Research Committee of the Chief of the General Staff considered a paper presented by the Inspector-General of the Royal Tank Corps which dealt with requirements for a tank to co-operate with infantry—the category which later became known as the "infantry tank". This postulated two alternatives: (1) a very small and inconspicuous vehicle, heavily armoured, armed with machine guns, and available in numbers; (2) a bigger vehicle with a larger calibre gun and heavier armour able to engage enemy weapons as well as carrying machine guns for defence against enemy infantry. In both cases the vehicle was only required to move at walking pace, ie, infantry speed.

The General Staff drew up a specification based loosely on the second alternative, though they asked for either a ·5in machine gun or a 2pdr as main armament, with a minimum armour thickness of 1in. Vickers designed the A9 (qv) to this specification. Major General Sir Hugh Elles was appointed Master General of the Ordnance in May 1934, responsible for tank procurement. Elles had commanded the Tank Corps in France in World War I and as a result of his experience was a keen advocate of tanks for infantry support. He was persuaded that Vickers could design and build a prototype infantry tank to the "small" specification but with increased armour, proof against any known calibre of anti-tank gun. Sir John Carden designed this vehicle, designated A11, and the pilot model A11E1, was delivered to the army for trials in September 1936.

To keep costs to a minimum, the A11 was very simple. A commercial Ford V-8 engine and transmission were used, with steering, brake and clutches adapted from the type used in the Vickers light tanks. The simple suspension was derived from that used on the Vickers Dragon gun tractors. Trials revealed the need for suspension improvements to prevent track shedding (the return rollers as a result being resited on the hull sides) and the need for a certain amount of splash proofing. A first production order for 60 vehicles was placed in April 1937 and this was later increased, a total of 140 vehicles (including the pilot model) being completed by August 1940 when production ceased.

Meanwhile, the limitations of machine gun arament were appreciated and design of the A12, Infantry Tank Mk II (qv) was initiated in November 1936, the A11 being regarded as an interim type. The Infantry Tank Mk I equipped the 1st Army Tank Brigade in France 1940 and proved almost immune to German anti-tank guns. However, its lack of hitting power rendered it of limited tactical value and production was abandoned after the Dunkirk evacuation. Remaining vehicles were used for training only.

SPECIFICATION

Designation: Tank, Infantry, Mk I, Matilda I (A11)
Crew: 2 (commander-gunner, driver)
Battle weight: 24,640lb
Dimensions: Length 15ft 11in; Height 6ft 1½in; Width 7ft 6in; Track width 11½in; Track centres/tread 6ft 4in
Armament: Main: 1 × ·5 cal Vickers MG or 1 × ·303 cal Vickers MG
Secondary: —
Armour thickness: Maximum 60mm; Minimum 10mm
Traverse: 360°. Elevation limits: —
Engine: Ford V8 70hp
Maximum speed: 8mph
Maximum cross-country speed: 5·6mph
Suspension type: Box bogie and leaf spring
Road radius: 80 miles
Fording depth: 3ft (short distance only)
Vertical obstacle: 2ft 1in
Trench crossing: 7ft
Ammunition stowage: 4,000 rounds
Special features/remarks: All riveted construction with cast turret. Name Matilda was bestowed by General Elles due to the vehicle's diminutive size and duck-like shape and gait.

98. The A11E1 pilot model. Note different nose layout and original form of return rollers.

99. Infantry Tank Mk I with Coulter Plough device for mine clearing.

VARIANTS

Infantry Tank Mk I with Coulter Plough: The problem of clearing paths for tanks through minefields was appreciated as early as January 1937 and two devices, the Coulter Plough and Fowler Rollers, were designed for fitting to the Infantry Tank Mk I. The Plough was eventually chosen and tested on a production vehicle in 1939. It consisted of two pivoted girder arms, mounted on the hull sides and raised and lowered via chains from a power take-off on the rear drive shaft. The device was never used in combat though provided experience for later adaptations of other vehicles.

100. Matilda I in use for crew training, late 1940.

101. Matilda I; note the stowage boxes each side of nose.

102. Infantry Tank Mk 11A*, Matilda III.

AT the time the Infantry Tank Mk I pilot model—A11E1—was delivered, the Mechanisation Board was already considering a "scaled up" version with an extra crew member, a 2pdr gun or twin machine guns, and a speed of up to 15mph. Weight limit of 14 (long) tons was imposed to meet current military bridging restrictions.

It soon became obvious that these requirements could not be incorporated in the basic A11 design since the turret alone, to hold a 2pdr gun, would bring the A11 weight up to 13 tons, and a new engine would also be required. Thus a completely new design was called for and was drawn up by the Mechanisation Board on the basis of 60mm armour thickness, a commercial type AEC diesel engine, and heavy side skirts to protect the "Japanese Type" suspension derived from that on the Vickers Medium Tank. The layout of the "Matilda Senior" as it was called, designated A12 Infantry Tank Mk II, was based closely on that of the A7 medium tank which had been designed and built in prototype form only by Royal Ordnance Factory, Woolwich, in 1929-32.

In November 1936 Vulcan Foundry of Warrington were given contracts to produce wooden mock-ups and two mild steel pilot models of the A12 design. The mock-up was inspected in April 1937, by which time it had been decided to use twin AEC diesel engines coupled together and a Wilson epicyclic gearbox. Provision was to be made for mounting a 3in howitzer in close support models and various other detail points were settled at this early stage. Construction of the pilot model was, however, held up by delays in the delivery of the gearbox and other components and the A12E1 pilot was not ready until April 1938. Meanwhile an order "off the drawing board" for 65 vehicles was given in December 1937, soon increased to 165. Trials were generally satisfactory, but some small modifications were made to the gearbox and suspension. Cooling was also improved and provision made for "colonial" use by fitting air cleaners.

By this time re-armament was under way and the need for tanks was urgent. In June 1938 contracts for further vehicles were placed with Fowler and Ruston & Hornsby under Vulcan's "parentage", and subsequently LMS, Harland & Wolff, and North British Locomotive all received contracts. For the later marks Leyland were brought in (in 1940) to make engines. Total output of A12s was 2,987 and production ceased in August 1943.

The A12 did not lend itself to easy mass-production, however, due to the size and shape of the armour castings used in the design. There was particular difficulty in making the one-piece armour side skirts and the number of mud chutes was reduced from six (in the pilot model) to five (in production models) to facilitate producing this component. By the outbreak of war with Germany in September 1939 there were only two A12s in service, though a number had been issued to 7th Royal Tank Regiment in France by early 1940 where they were used with success in the Battle of Arras just prior to the Dunkirk evacuation.

With withdrawal of the A11, Matilda I, the terms "Matilda Senior" and "Matilda II" were dropped as descriptions of the A12, and it was known simply as the Matilda. The Matilda is best remembered for its important part in the early Western Desert campaigns. In Libya in 1940 it was virtually immune to any Italian anti-tank gun or tank, and Matildas reigned supreme until the appearance of the German 88mm Flak gun in the anti-tank role in mid-1941, the first gun able to penetrate Matilda's heavy armour at long range. It was not possible to fit the 6pdr gun in the Matilda (though an attempt was made to mount the A27 type turret on a Matilda chassis), due to the small size of the turret and turret ring. Thus in 1942, the Matilda declined in importance as a gun tank and was last used in action in this role at the first Alamein battle in July 1942.

From then on, the Matilda was used in secondary roles for special purposes. Most important was its development as a mine-clearing vehicle, the extensive minefields laid by both sides in the desert fighting giving rise to its employment in this role. Major Du Toit, an engineer officer with the South African forces suggested a chain (or flail) device to beat the ground ahead of a slowly moving tank, thus detonating any mines in its path. He was sent by 8th Army HQ to Britain in November 1941 to work on this device. A pilot model of the type of vehicle he suggested was built by AEC under the auspices of the Ministry of Supply in December 1941. Known as the Baron Mk I, this was a Matilda tank retaining its turret and main armament and carrying a shaft and rotor about 10ft ahead of the vehicle, 6ft above the ground. A Chrysler engine drove the flail rotor and lift for the flail assembly was taken from the tank's hydraulic turret traverse system.

Both the Chrysler engine and the hydraulic system proved under-powered for the task when trials were conducted in January 1942. The vehicle was thus re-built with a 6 cylinder Bedford engine replacing the Chrysler, and hydraulic rams for the flail arms. Ready in April 1942, this modified vehicle was called Baron II. Trials, completed in June 1942, indicated the need for yet more powerful flail drive, a lower rotor height to give a "flatter" flail strike, and better cooling to prevent over-heating. Further design work resulted in a lower "perambulator" flail and the use of two Bedford engines to drive the rotor. These were mounted in armoured boxes at the rear of the vehicle with an operator to control them. This was the Baron III, but before work was completed, in September 1942, it was realised that the extra weight involved would adversely affect vehicle performance. Further modifications were thus made, resulting in removal of the turret and the substitution of a cab for the flail operator in the turret space. Known now as the Baron IIIA, the vehicle was ready for trials by the end of 1942. Flailing speed was ½mph. Production vehicles were ready by mid-1943, but by now the Scorpion device (see below), a much superior idea, was in production and work on the Sherman Crab was in hand. The Baron was therefore used only for training.

Du Toit's idea had meanwhile been taken up independently by Middle East Forces who produced the Scorpion I for fitting to the basic Matilda (though modified versions were also fitted later to the Grant M3 and Valentine tanks). This was much simpler than the Baron with fixed rotor arms and a single 30hp Bedford engine mounted on the right side of the hull—complete with operator's position—to drive the rotor. Thirty-two Scorpion-fitted Matildas were used at Alamein in October 1932. In December 1942, details were submitted to the War Office and Scorpion devices were put into production in Britain from February 1943 for fitting to the Valentine (qv).

Other special purpose and experimental variants of the Matilda are described below. Matilda gun tanks were used in the Eritrea campaign and also by the Australian Army in New Guinea. Some were still used by Australian reserve units in post-war years.

VARIANTS

Matilda I, Infantry Tank Mk II: First production type with AEC diesel engines, 2pdr gun and Vickers co-axial machine gun. Non-tropicalised engine in early vehicles.

Matilda II, Infantry Tank Mk IIA: As above but with Besa 7·92mm machine gun replacing Vickers.

Matilda III, Infantry Tank Mk IIA *: As above but with Leyland diesels replacing the AEC type.

Matilda III CS: As above but with 3in howitzer replacing 2pdr.

Matilda IV, Infantry Tank Mk IIA:** As Mk III but with improved Leyland engine.

Matilda V: As Mk IV but with improvements to gear box and gear shift.

Matilda II CDL and Matilda V CDL: The CDL (canal defence light) was an armoured housing with powerful searchlight fitted in place of the tank's original turret to illuminate the battlefield in night actions. This idea had been advocated in the thirties by a group of private individuals, and was demonstrated and sold to the War Office in 1937. In September 1939, an improved armoured turret was designed and after trials in 1940, 300 were ordered for fitting to Matildas—enough to equip one Brigade in Britain and one in the Middle East. Despite intensive training, CDLs were never used in their intended role until the Rhine crossing in 1945, mainly because suitable opportunities never arose. By 1945, however, the Grant (qv) had replaced the Matilda as the standard CDL-fitted tank.

Baron I, II, III and IIIA: Mine-clearing versions of the Matilda developed in Britain, 1942-43. For full description see main text.

Matilda Scorpion I: Mine-clearing device fitted to Matilda. Developed in Middle East. See text.

Matilda with AMRA Mk Ia: Another mine-clearing device, this consisted of Fowler rollers on a heavy frame pushed in front of the tank to detonate mines by pressure (AMRA: anti-mine roller attachment). In service in small numbers, some in Western Desert. Also fitted to other tanks (eg, Churchill).

Matilda with Carrot: Carrot demolition charge (600lb HE) carried on front of AMRA and detonated remotely from inside tank. Used for blowing gaps in obstacles. Light Carrot was a smaller charge similarly carried, though rollers were then removed from AMRA frame.

Matilda Frog: Australian flame-throwing device. 25 vehicles converted in late 1944. Tube replaced 2pdr gun in Matilda IV or V. Fuel carried in turret. Range up to 100yds. Used in New Guinea. Operated by gas pressure with 20 second delay between shots while pressure was recovered by pump.

Matilda Murray: Improved flame-throwing device developed in Australia to replace the Frog, which was limited tactically by the 20 second delay between shots. The Murray, produced in 1945, was similar in all respects to the Frog except that a cordite charge replaced gas pressure as a means of operation, thus allowing continuous bursts of flame.

Matilda Dozer: Australian development with box-shaped blade raised and lowered from vehicle's turret traverse hydraulic system.

Matilda with Inglis Bridge: Light bridge on tracked dumb carrier pushed ahead of adapted Matilda tank to span gaps under fire. Experimental only, but used for training in 1942. It led to the development of later bridging equipments of similar type evolved for the Churchill. The Inglis Bridge was of World War I origin, but remained in use in the earlier part of World War II.

Matilda with Trench Crossing Device: Another experimental device pushed ahead of a Matilda on light tracked bogies for spanning gaps to allow infantry and "B" vehicles to cross.

MATILDA II

103. A12E1 pilot model. Note the six mud chutes in side skirts.

104. Matilda Mk I (A12). Note armoured sleeve over co-axial Vickers MG.

SPECIFICATION

Designation: Tank, Infantry, Mk II, Matilda (A12)
Crew: 4 (commander, gunner, loader, driver)
Battle weight: 59,360lb
Dimensions: Length 18ft 5in; Height 8ft 3in; Width 8ft 6in; Track width 14in; Track centres/tread 6ft 9½in
Armament: Main: 1 × 2pdr OQF
Secondary: 1 × 7·92 cal Besa MG (Vickers MG in Mk 1)
Armour thickness: Maximum 78mm
Minimum 13mm
Traverse: 360°. Elevation limits: —
Engine: Twin Leyland 6 cylinder, 95hp each
Twin AEC diesels, 87hp each (Mks I–II)
Maximum speed: 15mph
Maximum cross-country speed: 8mph
Suspension type: Bell crank ("Japanese Type")
Road radius: 160 miles
Fording depth: 3ft
Vertical obstacle: 2ft
Trench crossing: 7ft
Ammunition stowage: 93 rounds 2pdr
2,925 rounds 7·92 cal MG
Special features/remarks: Easily recognised by heavy armoured side skirts concealing suspension. Bolted construction with cast nose and turret. Most heavily armoured British tank in service, 1940. First 10 were training vehicles built in mild steel and designated with "M" suffix.

105. Matilda CDL. Armoured light housing replaces turret.

106. Baron IIIA. Final form of the Baron development with turret replaced by flail operator's cabin. *Right*, flail in action; *below*, at rest.

107. Matilda IV CS showing the 3in howitzer carried in this version.

108. Matilda Scorpion. Note operator's armoured compartment on side and station keeping lights at rear for following vehicles.

109. Matilda with AMRA Mk I on test in Western Desert.

110. Matilda with Heavy Carrot device.

111. Matilda Frog flame-thrower in action.

112. Matilda Dozer in New Guinea.

113. Matilda propelling Inglis Bridge.

114. Matilda with experimental Trench Crossing Device.

115. Standard production Valentine XI; note 75mm gun.

EARLY in 1938, Vickers were among the firms approached to join the Infantry Tank Mk II (A12) production group under Vulcan. As an alternative they were invited to build a design of their own based on the A10 (qv) which had been the first "infantry tank" to emanate from the 1934 General Staff specification for this type (described in the Infantry Tank Mk I section). The A10 had subsequently been reclassified as a "heavy cruiser" since it was much less heavily armoured than the A11 and A12 designs. Vickers chose the latter alternative since they already had production facilities and experience for an A10-based design which would have been wasted if they had switched to building A12s. The new vehicle utilised a chassis, suspension, engine and transmission identical to the A10 but had a lower, more heavily armoured superstructure and an entirely new turret mounting a 2pdr gun. Plans were very quickly drawn up and submitted to the War Office just prior to St Valentine's Day in February 1938, and this date suggested the name "Valentine" by which the vehicle was subsequently known.

More than a year passed before a production order was placed, however, the main shortcoming in the design in the view of the General Staff being the small turret which would only accommodate two men. In July 1939, with war almost inevitable and an urgent need for tanks in quantity, an order for 275 vehicles was placed with Vickers straight off the drawing board. Quick production had been promised by Vickers since the chassis had already been proven with the A10, so that no lengthy development period was required. The first production vehicle was delivered to the Army for trials in May 1940 and proved very satisfactory, both a stable gun platform and mechanically reliable. First service deliveries were made in late 1940 and for a period in 1940-41 Valentines were used in the cruiser tank role in armoured divisions to help overcome the shortage of cruiser tanks. The first Valentines appeared with tank brigades in the 8th Army in June 1941 and subsequently played an important part in the remainder of the desert fighting.

Valentine production ceased in early 1944 after a total of 8,275 vehicles had been completed. By late 1942, however, the Valentine was largely obsolete due to its low speed and small turret which restricted the fitting of larger calibre armament. Mks III and V had the turret modified to accommodate three men (a loader in addition to commander and gunner), but the third turret member was, of necessity, dropped when the 6pdr gun was fitted in later marks. In this case the inadequacy of a two man turret crew was accepted in the interests of increased gun power. The 6pdr gun was introduced into production Valentines from March 1942. Other modifications included an improved engine installation (a GMC diesel unit) and a change over from all riveted to all welded construction. In March 1943 a Valentine was used for the test installation and firing of the British 75mm tank gun, intended for the A27 cruiser tanks and the Churchill, and the success of these trials in the Valentine led to the development of a production version mounting this gun. This version (Mk XI) was the final production variant. Characteristics of the various marks are given below. Valentines were built by Metropolitan-Cammell and Birmingham Carriage & Wagon in addition to Vickers.

Valentines were also built in Canada by the Canadian Pacific in Montreal. Of the 1,420 vehicles produced, however, all but 30—which were retained for training—were delivered to the Soviet Army.

The Valentine was one of the most important British tanks and in 1943 totalled nearly one quarter of British tank output. Valentines formed the basis of many special purpose AFVs including bridgelayers, mine clearers, and amphibious (DD) tanks. These are detailed below. Two SP guns were also put into production on the Valentine chassis, and these are described separately.

VARIANTS

Valentine I, Infantry Tank Mk III: Original production model with AEC 6 cylinder 135hp gasoline engine as in A10. Armed with 2pdr gun and co-axial Besa MG.

Valentine II, Infantry Tank Mk III*: As Mk I but with AEC diesel engine (131hp) replacing petrol engine. Vehicles for desert service fitted with sand shields and jettisonable long range fuel tanks.

Valentine III: As Mk II but with turret modified to take a third crew member. New turret front with improved mantlet; commander's hatch set towards rear of turret.

Valentine IV: As Mk II but with GMC diesel engine (138hp) replacing the AEC unit.

Valentine V: As Mk III with three-man turret but with GMC diesel engine in place of AEC unit.

Valentine VI: First Canadian production model, built to US/Canadian engineering standards but externally similar to Valentine IV. GMC diesel engine. In production late 1941. Nose plates cast instead of bolted as on British vehicles. Browning ·30 cal machine gun replaced Besa MG after sixteenth vehicle.

Valentine VII: Improved Mk VI built in Canada with different radio set and detailed internal changes.

Valentine VIIA: Improved Mk VII built in Canada with jettisonable fuel tanks, studded tracks, and protective cages over headlamps.

Valentine VIII: Mk III up-gunned with 6pdr replacing the 2pdr, and co-axial machine gun deleted. Turret crew reduced to two.

Valentine IX: Mk V up-gunned with 6pdr replacing the 2pdr and co-axial machine gun deleted. Turret crew reduced to two.

Valentine X: Vehicle with 6pdr gun and co-axial Besa MG as built. Up-rated GMC diesel engine (165hp). In production 1943.

Valentine XI: As Mk X but with 75mm gun replacing the 6pdr. Final production type, late 1943. Mks X and XI had splash plates protecting turret ring. Welded construction.

Valentine CDL: Vehicle with turret replaced by light armoured housing as Matilda CDL (qv). Few only converted.

Valentine OP/Command: Converted gun tank with dummy gun and extra communications equipment for battery commanders and OP officers of Archer-equipped SP units, 1944.

Valentine Scorpion II: Valentine tank with turret removed and replaced with armoured cabin for flail operator and rotor drive engine. This conversion was produced in Britain in mid 1943 and the Scorpion equipment was derived from that developed in the Middle East for fitting to the Matilda. Not used operationally since it was replaced from late 1943 by the Sherman Crab (qv). It was used for training however. For full description of Scorpion equipment, see Matilda Scorpion. This vehicle could also tow Centipede anti-mine rollers. A large counterweight was fitted at the hull rear on the Valentine Scorpion to balance the weight of the rotor arms in front.

Valentine AMRA Mk Ib: Valentine gun tank adapted to propel anti-mine roller attachment (AMRA). Not used operationally.

Valentine Snake: Valentine rigged to tow Snake equipment —an explosive-filled pipe towed across a suspected minefield then detonated to blow a passage through the mines. A few were so used in 8th Army.

Valentine Bridgelayer: Valentine II with turret removed and adapted to carry No 1 30ft scissors bridge. Hydraulic rams and arms fitted for launch and recovery. Hydraulic equipment fitted in turret space. Mainly used for training, 1943-44. Churchill Bridgelayer replaced it for operational use, but some were used in Burma. Bridge was 34ft × 9½ft, class 30.

Valentine Burmark: One prototype vehicle with twin Twaby Ark ramps fitted to Valentine chassis, 1945, for service in Burma—the Churchill Ark being too large for this area. Cancelled in 1946.

Valentine 7·92in flame mortar: Experimental vehicle consisting of Valentine with turret removed and replaced by a fixed heavy mortar intended to project 25lb TNT incendiary shells to demolish concrete emplacements. Trials only by Petroleum Warfare Dept, 1943-45. Maximum range of this weapon was 2,000 yards and effective range was 400 yards.

Valentine with 6pdr anti-tank mounting: Experimental vehicle built by Vickers to examine possibility of producing a simple tank destroyer by mounting 6pdr field carriage on hull in place of turret. Trials only, 1942.

Valentine DD (Mks III and VIII): The DD (Duplex Drive) idea was evolved by Nicholas Straussler, a Hungarian-born military engineer, and the system involved a propeller driven by power take-off from the vehicle's engine. To provide flotation, collapsible canvas side screens were fitted to the vehicle's hull which, when raised, gave a boat-like form to the vehicle. The actual tank hull was, in fact, suspended below the water surface. The DD principle was tested on a Tetrarch light tank (qv) at Brent Reservoir in June 1941 and proved successful. Subsequently the Valentine was selected as the standard DD tank, design was finalised in June 1942, and 650 Mks III and VIII were converted accordingly. They were mainly used for crew training and for developing operational techniques. In late 1943 the Sherman was similarly adapted and the DD brigades of 79th Armoured Division which took part in Operation Overlord and subsequent operations in NW Europe were equipped with the latter type.

Valentine Flamethrowers: To determine the best system for a tank-mounted flame projector, the Petroleum Warfare Dept (formed in June 1940) modified two Valentine tanks in 1941, one with a projector ignited by cordite charges and one with a projector operated by gas (nitrogen) pressure. The fuel was carried in a trailer and the flame projector was mounted on the hull front. Trials started in 1942 showed the gas-operated system to be the best, and from this test installation was developed the Crocodile equipment for the Churchill flamethrower used in the NW Europe campaign in 1944-45. The trailer of the Crocodile was essentially the same as that used with the Valentine, with the addition of an armoured superstructure. Neither Valentine flamethrower variant was used operationally.

Roller Fascine: 60 of these were built in 1940, made of 6ft 4in twin cable drums on a common spindle, which were designed to be pushed ahead of Valentine or Matilda tanks to assist in spanning anti-tank ditches. They were never used.

VALENTINE

SPECIFICATION

Designation: Tank, Infantry, Mk III, Valentine
Crew: 3–4 (commander, gunner, driver)
(plus loader in Mks III and V)
Battle weight: 39,000lb (41,000lb Mk VIII–XI)
Dimensions: Length 17ft 9in (19ft 4in, Mks VIII–XI); Track width 14in
Height 7ft 5½in; Track centres/tread 7ft 3in
Width 8ft 7½in
Armament: Main: See details for each mark in variants list.
Secondary: See details above.

Armour thickness: Maximum 65mm
Minimum 8mm
Traverse: 360°. Elevation limits: +20° to −5°
Engine: See details for each mark in variants list
Maximum speed: 15mph
Maximum cross-country speed: 8mph (approx)
Suspension type: "Slow motion" with 3-wheel bogies
Road radius: 90 miles
Fording depth: 3ft
Vertical obstacle: 2ft 9in
Trench crossing: 7ft 6in
Ammunition stowage: 79 rounds (2pdr versions),
3,150 rounds 7·92 cal MG
53 rounds (6pdr versions),
1,575 rounds 7·92 cal MG
Special features/remarks: No ordnance designation ("A" number) allocated since this vehicle was not designed to match a General Staff specification. Electric or hand turret traverse; manganese tracks; most reliable of British tank designs of early war years.

116. First production vehicle, Valentine I.

117. Valentine II fitted with sand shields and auxiliary fuel tanks for Middle East service.

118. Valentine III with three-man turret. Note later pattern mantlet used in this and subsequent marks with 2pdr gun.

119. Valentine IX with 6pdr gun. Note new mantlet.

120. Valentine Scorpion. Note counterweight at hull rear.

121. Valentine with AMRA Mk Ib. These were known as Fowler rollers.

122. Valentine II with Snake equipment.

123. Valentine Burmark showing front ramps being launched.

124. Valentine Flamethrower with cordite-operated equipment.

125. Valentine Flamethrower with gas-operated equipment.

126. Valentine III and VIII (at rear) DD tanks with side screens lowered. Note extended exhaust pipes, and propellers.

127. Valentine with experimental 7·92in flame mortar.

128. Valentine with experimental 6pdr anti-tank gun mount.

129. Valentine Bridgelayer, bridge in travelling position.

130. Valentine Bridgelayer, showing bridge detail.

131. Bishop SP 25pdr in the Western Desert.

THE successful use by the Germans in the Western Desert of self-propelled guns as infantry support weapons led to a request from the 8th Army HQ for the urgent provision of similar equipment. A scheme for mounting the 25pdr gun-howitzer on an existing tank chassis was suggested and Birmingham Carriage & Wagon were asked in June 1941 to give priority to developing such a vehicle utilising a Valentine tank chassis. A pilot model was quickly produced and was ready for trials by August 1941. The conversion was essentially simple—basically the 25pdr weapon mounted in a fixed turret with a box-like shield to protect the crew. Trials were successful but several changes were requested, mainly to increase crew protection.

An order for 100 vehicles was given in November 1941 with the promise of a further 200 to follow. However, in March 1942 the British Tank Mission in USA, responsible for procuring American tanks for British service, saw and ordered the M7 Howitzer Motor Carriage (qv) an SP vehicle based on the M3 medium tank chassis. This was a much superior vehicle to the Valentine with 25pdr and further orders for this vehicle were cancelled. In July 1942, when British fortunes in the desert fighting were at their lowest ebb, a new order for 50 Valentine SPs was placed. At this period 80 of the first 100 had been delivered and a number had been shipped to the 8th Army. The small number of Valentine SPs in the desert were soon supplemented by the M7 which began to enter service with the British in July 1942 in large numbers. Named the Bishop the Valentine SP was in use until the end of the North African campaign in 1943, after which it was used mainly for training. Relatively unimportant, the Bishop was the first British designed SP vehicle of World War II. Due to the initial indecision about getting it into production—a year passed from the first request to delivery—this vehicle was eclipsed by a superior replacement, the M7 Priest, almost as soon as it was ready. Compared to the M7 Priest, the Bishop was a crude and unsophisticated extemporisation.

SPECIFICATION

Designation: Carrier, Valentine, 25pdr gun Mk I, Bishop
Crew: 4 (commander, driver, gunner, loader)
Battle weight: 39,000lb
Power/weight ratio: 6·6HP/ton
Dimensions: Length 18ft 2in; Height 9ft 3¼in; Width 8ft 7½in; Track width 14in; Track centres/tread 7ft 3in
Armament: Main: 1 × 25pdr gun howitzer Mk I
Secondary: 1 × ·303 cal Bren MG (AA)
Armour thickness: Maximum 60mm
Minimum 8mm
Traverse: 4° left, 4° right. Elevation limits: +15° to −5°
Engine: AEC diesel 131hp
Maximum speed: 15mph
Maximum cross-country speed: 7mph (approx)
Suspension type: "Slow motion" with 3-wheel bogies
Road radius: 90 miles
Fording depth: 3ft
Vertical obstacle: 2ft 9in
Trench crossing: 7ft 6in
Ammunition stowage: 32 rounds, 25pdr
Special features/remarks: Built on Valentine II chassis. High, square, open-topped shield for gun mounting. Very high silhouette was a disadvantage for desert fighting. Sometimes towed a standard wheeled ammunition limber carrying extra 25pdr rounds.

132. Standard production 17pdr SP, Archer, side view.

DESIGN of the 17pdr gun as a high velocity anti-tank weapon comparable in hitting power to the German 88mm gun began in the fall of 1941. It was approved for production in mid 1942 and consideration was given to fitting it in tanks (giving rise to the Challenger and Sherman Firefly) and in self-propelled mounts for the "tank destroyer" role. Of existing tanks which could be adapted, the Crusader was ruled out as being too small and underpowered to take the mounting, leaving the Valentine as the only available alternative existing in quantity. The earliest idea for mounting the 17pdr on this chassis was to use the Bishop, already in production, with the new weapon substituted for the 25pdr gun-howitzer. This was impracticable due to the length of the 17pdr barrel and the already unwieldy height of the superstructure. The Ministry of Supply therefore asked Vickers to design an entirely new SP vehicle based on the well-proven Valentine chassis but overcoming the limitations imposed by the extreme size of the weapon. Work on this started in July 1942 and the pilot model was ready for trials in March 1943.

The new vehicle, which was named Archer, was basically a Valentine chassis with an open topped superstructure over the fighting compartment. The 17pdr gun was mounted to point to the rear with limited traverse. The driver was in the same position as in the original tank and the superstructure front was virtually an extension of the glacis plate. Thus, despite the length of the 17pdr barrel, a reasonably compact, low SP vehicle was produced. Firing trials were carried out in April 1943, resulting in requests for detail changes to the mount and fire control equipment. On the whole the vehicle was satisfactory and was placed in priority production. The first production model was completed in March 1944, and the Archer equipped anti-tank battalions of British armoured divisions in NW Europe from October 1944 onwards. It remained in service until the mid fifties with the British Army, and Archers were also supplied to several other armies in post-war years. A total of 665 were built, all by Vickers, out of an original order for 800.

Despite the severe tactical limitation of the rear-pointing gun, the Archer—initially regarded as an interim type while better designs were worked out—proved a reliable and effective weapon.

133. Archer seen from rear; gun always pointed to rear.

SPECIFICATION

Designation: SP 17pdr, Valentine, Mk I, Archer
Crew: 4 (commander, gunner, loader, driver)
Battle weight: 36,960lb
Dimensions: Length 21ft 11¼in; Height 7ft 4½in; Width 8ft 7½in; Track width 14in; Track centres/tread 7ft 3in
Armament: Main: 1 × 17pdr OQF
Secondary: 1 × ·303 cal Bren MG (AA)
Armour thickness: Maximum 60mm
Minimum 8mm
Traverse: 11° right, 11° left. Elevation limits: +15° to —7½°
Engine: 1 × GMC diesel 165hp
Maximum speed: 15mph
Maximum cross-country speed: 8mph
Suspension type: Slow motion with 3-wheel bogies
Road radius: 90 miles
Fording depth: 3ft
Vertical obstacle: 2ft 9in
Trench crossing: 7ft 6in
Ammunition stowage: 39 rounds, 17pdr
Special features/remarks: Low open-topped superstructure at front with gun pointing to rear. Original order was for 800 vehicles but this was cut back to 665 in 1945. All-welded construction. For cross-section of this vehicle see plate 529.

134. Churchill I showing original appearance of the design. Note 3in howitzer in nose.

AT the outbreak of war in September 1939, there was a strong school of opinion at the War Office that conditions on the Western Front would not be very different from conditions experienced in 1914-18. There was therefore a need for a very heavy infantry tank invulnerable to known anti-tank guns, with a very wide trench-crossing ability, and able to negotiate ground churned up by shell fire. Designated A20, a specification was drawn up by the Superintendent of Tank Design, Woolwich, and Harland & Wolff were asked to build a pilot model. Armour thickness of 80mm, speed of 15mph, ability to climb a 5ft parapet, and a crew of seven were among characteristics requested. Essentially the A20 was a refinement of the "lozenge" shape tanks built by the British in 1916-18; various combinations of armament were considered including a 6pdr, French 75mm, 3in howitzer, and 2pdr. Finally a 2pdr was selected for the turret, another to be mounted in the nose, with machine guns recessed in the hull side at the front. Four pilot models were ordered in February 1940. The first pilot model ran trials in June 1940, plagued by gearbox trouble. Data from the first run, however, showed that in order to maintain the required performance the gun armament would have to be reduced to a single 2pdr.

This coincided with the Dunkirk evacuation, when Britain was left with less than 100 tanks for home defence. Vauxhall (the British off-shoot of GMC) were therefore asked to "refine" the A20 design, scale it down slightly, and get it into production as rapidly as possible, preferably within a year. Choice of Vauxhall was largely influenced by the fact that their Vauxhall-Bedford twin-six engine was scheduled for the A20. The A20 pilot model and plans were handed over to Vauxhall, extra draughtsmen were loaned by the Mechanisation Board, and a pilot model of the new design, A22, Infantry Tank Mk IV, was ready by November 1940. The first 14 production models were delivered in June 1941 from an order for 500 straight off the drawing board. Due to the rushed development programme for this vehicle, there were numerous defects in the design leading to frequent breakdowns with the early marks. This necessitated considerable re-work programmes in 1942-43, the secondment of Vauxhall engineers to units equipped with the tank, and numerous detail improvements to mechanical components.

Named Churchill, the A22 was built in quantity by a production group consisting of Broom & Wade, Birmingham Carriage & Wagon, Metropolitan Cammell, Charles Roberts, Newton Chambers, Gloucester Railway Carriage, Leyland, Dennis, and Harland & Wolff, all under the "parentage" of Vauxhall.

The Churchill was of composite construction consisting of an inner skin of $\frac{1}{2}$in mild steel with an outer covering of armour plate bolted or riveted in position. Initially a cast turret was fitted, but later models had larger turrets of either cast, welded, or composite construction. The engine and drive were at the rear, and the overall tracks with small sprung bogie assemblies allowed space between the lower and upper runs of track for stowage of ammunition and stores, making the Churchill an unusually roomy vehicle. Escape doors for the crew were fitted in each side. Transmission featured the new Merritt-Brown four-speed gearbox which provided controlled differential steering, the Churchill being the first British tank to have this.

Armament of the Mk I was a 2pdr with a 3in howitzer in the hull front. Changing tactical requirements, however, led to a change of armament through the Churchill's production life. In common with the British cruiser tanks a 6pdr gun was fitted in 1942, necessitating a larger turret (Mk III). Experience in the desert fighting of 1941-42 led the War office to believe that speed and reliability were more important than heavy armour, and it was decided to cease Churchill production in 1943 when the A27 series of cruiser tanks became available. However, the Churchill's first combat actions, with the 1st Army in the Tunisian campaign, proved most successful in the hilly conditions of the terrain and this earned the vehicle a reprieve. In 1943, the Churchill was again up-gunned (Mk VII) with the new British version of the 75mm gun. At the same time major design improvements were effected. Since it was built to meet British railroad loading guage restrictions, the Churchill suffered from the same disadvantage as other contemporary British designs in that it was too narrow to take the larger turret required for the 17pdr gun. Thus by 1944-45 it was under-gunned by German standards, but this was offset to an extent by the vehicle's heavy armour protection. Detail differences between various marks are outlined below.

The other factor which made the Churchill one of the most important British tanks of 1939-45 was its adaptability to the specialised armour roles needed for the invasion of Europe in 1944. The vehicle's roomy interior, regular shape, and heavy armour made it particularly useful as an armoured engineer vehicle, bridgelayer or recovery vehicle. The many variants built for these specialised roles are also detailed below. Finally there were many experimental variants of the Churchill and the most important of these are also outlined.

SPECIFICATION

Designation: Tank, Infantry, Mk IV (A22)
Crew: 5 (commander, gunner, loader, driver, co-driver-hull gunner)
Battle weight: 87,360lb (Mk III–VI), 89,600lb (Mk VII–VIII)
Dimensions: Length 24ft 5in; Track width 14in
Width 9ft (8ft 2in, Mks I–11); Track centres/tread 12ft 6in
Height 10ft 8in (11ft 4in, Mk VII–VIII)
Armament: Main: See variant details
Secondary: See variant details
Armour thickness: Maximum 102mm (152mm, Mk VII–VIII)
Minimum 16mm (25mm, Mk VII–VIII)
Traverse: 360°. Elevation limits: +20° to −12½°
Engine: Bedford twin-six 350hp
Maximum speed: 15½mph (12½mph, Mk VII–VIII)
Maximum cross-country speed: 8mph (approx)
Suspension type: Sprung bogies
Road radius: 90 miles
Fording depth: 3ft 4in (unprepared)
Vertical obstacle: 2ft 6in
Trench crossing: 10ft
Ammunition stowage: See variant details
Special features/remarks: Basic marks most easily recognised by type of turret and/or gun. Mk VII–VIII all-welded with integral armour; all others of composite construction. Lighter and faster version of Churchill, designated A26, was proposed as a heavy cruiser tank but did not pass project stage. Some very minor and unimportant Churchill experimental types are omitted from the variant details. Total Churchill production: 5,640 vehicles.

COMBAT VARIANTS

(1) GUN TANKS

Churchill I: Original production model, 1941. Cast turret with 2pdr gun and co-axial 7·92 cal Besa MG. 3in howitzer mounted in nose. 150 rounds of 2pdr, and 58 rounds 3in ammunition. Some Churchill Is were used in the Dieppe raid, August 19, 1942.

Churchill II: As Churchill I but with 3in howitzer replaced by a second Besa MG.

Churchill IICS: As Churchill I but with gun positions changed so that 3in howitzer was in turret and 2pdr in nose. Built only in small numbers.

All the above marks originally appeared with fully exposed tracks and engine intake louvres (on hull sides) which had

135. Churchill II showing Besa MG in hull and strengthened front horns.

136. Production Churchill III with track covers and the revised side intake louvres.

137. Churchill IV with the Mk 5 6pdr gun (note counterweight), towing monotrailer with extra fuel.

138. Churchill V was close-support version of Mk IV, fitted with 95mm howitzer.

139. Churchill VII; note welded/cast turret, circular escape doors, and 75mm gun.

side openings. From May 1942 a "rework" programme was started and most early vehicles were fitted with full track covers and had strengthening plates fitted in the front horns. In addition new air intake louvres were fitted which had the opening on the top to prevent engine flooding when wading. For deep wading, trunking could be fitted to the top of the air intakes. There were also, of course, mechanical improvements.

Churchill III: Appearing in March 1942, this mark was the first to mount a 6pdr gun in line with War Office policy of upgunning cruiser and infantry tanks (further details included with A27 cruiser tank entry). To take the 6pdr gun a new welded turret was introduced which considerably altered external appearance. Early Mk IIIs produced before May 1942 had the original type air intakes and lacked track covers. Many Mk Is and IIs were brought up to Mk III standard with new turret and gun.

Churchill IV: As Churchill III with 6pdr gun but with a new cast turret which offered armour protection slightly superior to the welded turret of the Mk III. First produced in mid 1942. Turrets apart, both the Churchill III and IV were identical. 84 rounds of 6pdr ammunition were carried. Most vehicles had the Mk 3 6pdr gun but some early production models had the Mk 5 6pdr, distinguished by its longer, lighter appearance and frequently seen with a counterweight on its muzzle.

Churchill IV (NA 75): The appearance of the American M3 and M4 Medium tanks in the desert emphasised the need for guns of 75mm calibre in British tanks with a "dual purpose" HE/AP capability. The workshops of the brigades equipped with Churchills in 1st Army in Tunisia in January 1943 acted on their own initiative and fitted 120 of their Mk IVs with complete M3 75mm guns and mantlets salvaged from wrecked Sherman tanks. These were the first Churchills (or indeed British tanks) to take guns of this calibre into action when the NA 75s were used with great success in the Sicily and Italian campaigns, remaining in service until 1945. In this form, 84 rounds of 75mm ammunition were carried. (NA: North Africa).

Gun Carrier, 3in, Mk I, Churchill: In September 1941, the General staff asked the Tank Board to investigate the possibility of producing cruiser and infantry tanks mounting large calibre high velocity guns specifically for engaging the largest German tanks. To fulfil the the cruiser tank requirement the Challenger (qv) was subsequently developed

140. Churchill VI was Mk IV brought up to near Mk VII standard with 75mm gun in place of 6pdr.

141. Churchill VIII was close-support version Mk VII, with 95mm howitzer.

142. Churchill IV (NA 75) was Mk IV fitted with 75mm gun and mantlet from American Sherman tank.

with the 17pdr gun. For the infantry tank requirement it was proposed to fit a 3in AA gun in a limited traverse mount on a Churchill chassis. These weapons were available having been replaced in AA units by the 3·7in gun. A hundred vehicles were provisionally envisaged, but in December 1941 it was decided that all Churchill production would be needed for gun tanks, fitted with the 6pdr then just available. The order was thus reduced to 24 vehicles only. The pilot model was produced in February 1942 but these vehicles were never used operationally. Most were converted to carry Snake mine-clearing equipment and used for trials and training with this in 1943-44. Designated A22D.

Churchill V: Produced concurrently with the Churchill IV, this was the same basic vehicle fitted with a 95mm howitzer in place of the 6pdr, for the close support role. Only 10% of Churchill output had the 95mm howitzer (including the Mk VIII—see below). 47 howitzer rounds were carried.

Churchill VI Experience in the desert fighting led to demands from tank men for a 75mm gun in British tanks comparable to the 75mm guns fitted in the American M3 and M4 Mediums which had proved so successful in 8th Army hands. A British version of the 75mm gun, able to fire American ammunition, was developed by Vickers (for fuller details see A27 Cromwell entry) and plans were put in hand in January 1943 to mount this in the Churchill. The 75mm gun was not available until the winter of 1943-44 and a new mark of Churchill was meanwhile designed specially to mount it, the Mk VII (see below). At the same time, from November 1943 onwards, existing Mk IVs were converted to roughly Mk VII standards by the addition of a cupola, vane sights, and the 75mm gun in place of the 6pdr. With the Mk VII, the converted vehicles, designated Mk VI, were used in NW Europe, 1944-45.

Churchill VII: This was a largely redesigned version of the Churchill with thicker integral armour (as opposed to the composite construction on earlier Churchills), a new cast/welded heavy turret with cupola, circular (instead of square) escape doors, heavier suspension, improved gearbox, the 75mm gun, and many other refinements. Designated A22F, it was used in the NW Europe campaign 1944-45, and for many years post-war. 84 rounds of ammunition were carried. In 1945 this vehicle was re-designated A42.

Churchill VIII: As Churchill VII but with 95mm howitzer replacing the 75mm gun (see also Mk V, above).

Churchill IX: This was the Churchill III or IV "re-worked" to improved standards by fitting the same cast/welded heavy turret as the Mk VII but retaining the 6pdr gun. Other improvements were the addition of appliqué armour on sides and front of the hull.

Churchill IX LT: Reworked Churchill III or IV as above but retaining the original turret (LT: Light Turret).

Churchill X: Mk VI reworked as Mk IX, but armed with the 75mm gun.

Churchill X LT: Reworked Mk VI retaining its original turret.

Churchill XI: Mk V reworked with heavy turret (as on Mk VIII), plus appliqué armour.

Churchill XI LT: Mk V reworked as for Mk XI, but retaining its original turret.

Churchill Oke: This was a flame-thrower tank developed by the Petroleum Warfare Dept very quickly in 1942 so that the idea of a flame-throwing vehicle could be tested under combat conditions in the Dieppe landing, August 1942. Basically a Churchill II, it had a complete Ronson flame-throwing system installed (which had been developed for use in the Universal Carrier). The flame fuel containers were mounted at the rear, and the pipe was passed through the left pannier to project between the front horns. Three vehicles took part in the Dieppe landing, but all were destroyed before they could be used. Range of the Oke was 40–50 yards.

143. The Churchill 3in Gun Carrier.

144. Churchill X LT showing appliqué armour on sides and nose. This is a reworked Mk VI retaining its original turret.

145. Churchill Oke; this is a Mk II with flame-throwing equipment. Note flame projector on right.

Churchill Crocodile: After trials with flame fuel carried in trailers and tested on Valentines (qv) in 1942, the General Staff decided to standardise on a flame-throwing system actuated by gas (nitrogen) pressure. Design was finalised in 1943 with an initial order for 250 units featuring armour protection for the fuel trailer, and the Churchill IV was selected as operating vehicle. In October 1943 the Mk VII was chosen instead. Fuel passed along the belly via a "link" with the trailer, and the projector replaced the hull machine gun. Late production Churchill VIIs were all built for speedy

146. Churchill Crocodile flame-thrower. Vehicle is a Mk VII used in all Crocodile combinations.

adaptation to the Crocodile role as required. Vehicle's main armament could still be used, of course. Range of the Crocodile was 80–120 yards in 80 one-second bursts from a full trailer. When empty or hit the trailer could be jettisoned. Trailer weighed 6½ (long) tons. Used in NW Europe 1944-45, total production was 800 Crocodile units by May 1945, 250 earmarked for the Far East.

Churchill AVRE: (AVRE: Armoured Vehicle, Royal Engineers) Though the Dieppe raid was unsuccessful it proved that specialised types of armour were needed to assist assault forces in landing on and taking fortified open beaches. Urgently shown to be necessary was a heavily armoured vehicle to carry and support assault engineers charged with breaching heavy defences. Lt. Donovan of the Royal Canadian Engineers proposed adapting an existing tank. Both the Ram and Sherman were evaluated for the role, but the choice fell on the Churchill which had a roomier hull plus side escape doors which were useful for egress under fire. A spigot mortar, called a Petard, of 29cm calibre and firing a 40lb bomb 80 yards, was developed, tested in a Covenanter tank, and modified for fitting on the 6pdr mount of the Churchill III or IV. 180 Churchills of these marks were converted to AVREs by D-Day, June 6, 1944, and equipped the 1st Assault Brigade of 79th Armoured Division at this time. Subsequently another 574 vehicles were converted and AVREs played an important part in the NW Europe campaign. AVREs were fitted to carry and drop fascines (brushwood bundles), the CIRD (Canadian Indestructible Roller Device) for mine-clearing, and SBG (Small Box Girder) bridges, attachment points being incorporated for all these to be handled as required. A few AVREs were unarmed or lacked the usual AVRE fittings, these being used mainly for training. Also produced for use with the AVRE was a sledge for towing stores, fascines, or explosives. Further AVRE development took place post-war, again with the Churchill as a basis, though this is beyond the scope of the present volume.

(2) Recovery Vehicles

Churchill ARV Mk I: Churchill I or II with turret removed, stores carried in turret space, and fitted to carry demountable A-frame jib front or rear. Twin mount for Bren AA machine guns. Produced from February 1942.

Churchill BARV: Vehicle converted as for ARV I (above) but with shingle plates fitted over suspension arms and deep wading gear installed. Some had a box shaped dummy turret. Only a few of these were produced for recovering tanks on beaches.

Churchill ARV Mk II: Churchill III or IV chassis with turret removed and replaced by fixed box-like dummy turret and dummy gun. Fitted with demountable jibs, front and rear, earth spade at rear, and two-speed winch with 25 (long) tons pull. Produced in 1944 and used for many years post-war.

(3) Bridging Vehicles

Churchill Ark Mk I: A further lesson learnt from the Dieppe raid was the need for armoured bridge-carrying vehicles able to lay ramps across sea walls or span defence ditches and craters so that following vehicles could cross. In the fall

147. Churchill AVRE (right) and Churchill AVRE carrying SBG bridge. Note sheerlegs and rear support frame.

148. Churchill AVRE (conversion from a Churchill IV) showing Petard mortar and attachment point for CIRD. Note spare bogie assembly on hull top.

149. Churchill ARV Mk I with A-frame jib erected. Note twin Bren MG mount.

150. Churchill ARV Mk II with front jib erected. Note earth spade at rear and dummy gun.

151. Churchill Ark Mk II (UK Pattern) with ramps in raised position.

152. Churchill Ark Mk II (Italian Pattern). Note absence of trackways on hull.

of 1943, 79th Armoured Division, which was responsible for developing special purpose armour for the invasion of France, built an experimental version of the Churchill with the turret removed, timber trackways above the hull top, and ramps supported by kingposts and hinged at each end of the trackways. This was successfully tested and 50 further conversions of this type were ordered in February 1944, using Churchill II and IV chassis which had originally been earmarked for CDL use. ("Ark": Armoured Ramp Carrier).

Churchill Ark Mk II (UK Pattern): In July 1944 modifications were made to an existing Ark which involved doubling the width of the left hand trackway from 2ft to 4ft, so allowing "B" vehicles with narrow track centres to use the Ark crossing facilities as well as tanks. This was successful and all Ark Is were converted to this form and designated Ark II. The suffix "UK Pattern" was added to distinguish this type from a similar design evolved by 8th Army in Italy (see below). Arks had their turret apertures plated in and a square cupola added. A kingpost at each end held the ramps upright while the vehicle was driven into the required position, either in a ditch, a river, or against a wall or bank. The ramps were dropped by a quick-release mechanism so forming a bridge for vehicles to cross. The Ark was considered expendable if need be and no provision was made for rehoisting the ramps, except by the aid of a recovery vehicle. Sections of the trackways could be removed for engine access. Arks had a crew of four. Standard ramp length was $12\frac{1}{2}$ft, but longer ramps were tested and, on occasion, used for bridging extra wide gaps. There was also a 9ft long extension ramp which could be hinged to the end of the standard ramps if needed for specific tasks.

Churchill Ark Mk II (Italian Pattern): This was similar to the UK Pattern (as above), but utilised US made ramps which came in two different lengths, 12ft $3\frac{1}{4}$in (M2) or 15ft 3in (M1) for fitting as required. Fundamental difference from the UK Pattern vehicle was the lack of built-up trackways on the hull top, the vehicle tracks themselves serving as trackways. Mode of operation was the same. These vehicles were conversions on Churchill III hulls by Army workshops in Italy. They were at first designated Octopus.

Churchill Bridgelayer: Development of this vehicle started in 1942 and was based on experience gained with the Covenanter and Valentine Bridgelayers (qv). Essentially this was a turretless Churchill III or IV (or VII from 1945-46) with hydraulic equipment fitted in the fighting compartment to work a pivoted arm which could launch the bridge (Bridge, Tank, 30ft, No 2) horizontally from its stowage on the hull top and into position spanning a ditch or crater. The bridge was made in four parts for ease of handling but was carried, and launched, rigid. It could support vehicles up to 60 tons. Production vehicles were

153. Churchill Bridgelayer with bridge in travelling position.

issued in 1944 and allocated at first in troops of three to Brigade HQ of Churchill-equipped tank brigades. Later, as more vehicles became available, they were issued more widely. Crew of this vehicle was two. The bridge weighed 4·8 (long) tons. From 1946 on, a heavier (No 3) bridge was used on this vehicle.

Churchill AVREs were also used operationally to propel various types of assault bridge used in combat. These included the Skid Bailey, a short bridge built from Bailey parts, mounted on skids, and pushed and pulled into place by two AVREs, and the Mobile Bailey Bridge which was a complete (Class 40) bridge mounted on dumb Orolo track units. Two AVREs were also used to propel this. Two other similar types were the Brown and Dalton Mobile Bridges,

154. Churchill Bridgelayer shown launching bridge from hull top.

155. Churchill Great Eastern Ramp; rear view with rear ramps in raised travelling position.

used in Italy, built from Bailey parts and pushed by one AVRE while a second AVRE, with turret removed, acted as the carrier vehicle.

Numerous experimental bridging devices were also evolved using the basic Churchill chassis. Most important of these were: **Lakeman Ark**—a standard Churchill with elevated trackways built above turret height and a sloping ramp at rear; it was intended for surmounting extra high walls. **Great Eastern Ramp**—A 1944 scheme for a vehicle similar to an Ark but with the trackway elevated and much heavier; the front (25ft) ramps were fired into position using rockets. The pilot model was successful and 10 trials vehicles were delivered for combat testing in 1945; by this time, however, requirement for such vehicles had lapsed and the project was dropped. **Woodlark**—This was similar to the UK Pattern Ark but had twin triple-hinged ramps which were launched by rocket; trials only. There were also post-war Churchill bridging developments, outside the scope of this book.

Experimental and Limited Production Variants

(1) Anti-Mine Devices

Numerous devices for lifting or exploding mines were produced for use with the Churchill. Some of them, like the AMRA and CIRD, were also used with other British tanks. The most important were:

Churchill with AMRA Mk IIe: The AMRA (anti-mine roller attachment) was evolved and built by J. Fowler & Co and, from 1937, they produced variations adapted for fitting to nearly all British operational tank types (see Matilda, Covenanter entries, etc). Popularly known as "Fowler Rollers", the mark number indicated the tank for which they were intended. The heavy rollers were mounted on a frame carried in front of the tank and were on castored sprung mounts, exploding mines by pressure. The entire AMRA could be jettisoned if more than one roller was blown away. Not used operationally, stocks of AMRA were nonetheless produced.

Churchill with AMRCR: Developed in 1943 from the AMRA, the AMRCR (anti-mine reconnaissance castor roller) was more sophisticated in that it had a more flexible mounting for the roller and four larger rollers in place of the six on the AMRA. The whole device could be jettisoned from within the tank. Produced in small numbers only. A Mk 1a version was designed for fitting to the Sherman V.

Churchill with CIRD: The CIRD (Canadian indestructible roller device) was designed and produced by the Canadian Army in Britain in 1943. Two arms attached to the front of the tank each supported a heavy roller. The roller was made flexible enough to jump in the air and rotate in an arc round the arm when a mine was detonated, thus reducing the chance of the rollers being blown off. The CIRD could also be fitted to the Churchill AVRE, but was rarely used. Various roller sizes were produced from $15\frac{1}{2}$–21in diameter.

Churchill with Ploughs A-D: These were various combinations of ordinary agricultural ploughshare mounted on arms carried ahead of the tank. Rollers supported the arms. The idea with this equipment was to dig up mines and throw them clear of the path of the tank. Plough D, used with Atherton Equipment, was heavier than the rest. All featured a gantry projecting aft of the tank to counterweight the arms in front of the vehicle. These were experimental only.

156. Churchill AVRE with CIRD fitted (16in wheel pattern).

157. Unarmed Churchill AVRE with Plough A equipment in raised position. CIRD attachment point is used.

158. Churchill with AMRCR equipment. For AMRA appearance, also fitted to Churchill, see Matilda AMRA picture.

159. Unarmed Churchill AVRE trials vehicle with Jeffries Plough in raised position.

Churchill with Bullshorn/Jeffries Ploughs: These were two very similar devices, evolved by 79th Armoured Division, and consisting of ploughshares on light girder frames supported ahead of the vehicle just clear of the ground. Simplest of all "plough" devices, the Bullshorn was one of the few used operationally in the Normandy landings, June 1944.

Farmer Ploughs: These were developed in 1943 for use with the same attachment points as supplied for the AMRCR

160. Churchill IV with Farmer Front plough equipment. Note how CIRD attachment point is used.

161. Rare but interesting view of the early Churchill TLC Carpet Laying Device first used in the Dieppe raid, August 1942.

162. Churchill AVRE with Bobbin Mk II commencing to lay.

163. Churchill Mk IV with TLC Carpet Laying Device. Note release cord to bobbin. This is the type with canvas mat.

equipment, the object being to achieve standardisation. Three variants: Farmer Front, Farmer Deck, Farmer Track, all varying in detail with various combinations of tines and ploughshares. Not used operationally.

(2) Mat Layers

Developed for laying a trackway over marshy ground or barbed wire for wheeled vehicles and infantry, a rudimentary form of trackway on a bobbin attached to the front of a Churchill I was tried out in the Dieppe raid over the shingle beach. As a result of this 79th Armoured Division evolved more sophisticated forms of canvas trackway on bobbins. A beach at Brancaster, Norfolk with similar characteristics to the Normandy beaches, was used for trials in 1943. Types evolved were:

AVRE with Bobbin Mk I and Mk II: A canvas mat 9ft 11in wide on a spindle supported on short arms above the front horns. The Mk II had fixed arms; the Mk I had movable arms.

AVRE with Log Carpet Device: This consisted of a carpet of 100 6in diameter logs, each 14ft long, bound together with wire rope. A removable steel frame was fitted above the AVRE superstructure carrying the looped mat which was released over the front of the vehicle by detonating a light charge.

AVRE with Twin Bobbins: Two bobbins, one of canvas and one of chespaling carried ahead of the vehicle on horizontal front arms, allowing a choice of mats to suit the surface. Experimental only.

TLC Laying Devices: (TLC: tank landing craft) These predated the Bobbins, being developed in 1940 with a view to mounting on the Cruiser Mk I (A9) and Matilda II tanks. They were designed to be laid across a beach from a landing craft to prevent vehicles bogging in sand or shingle, but

164. Churchill Gun Carrier converted to carry Snake equipment.

165. Churchill AVRE towing Conger equipment, transported in an engineless Universal Carrier.

166. Ardeer Aggie; a converted Churchill III.

could also be used inland. They were developed for fitting to the Churchill in 1942 and the first type consisted of a canvas mat on front mounted arms while the second was a chespaling mat. The end was weighted and fell to the ground on release. The vehicle then ran over the end as it moved forward, thus pulling the rest of the mat from the spindle. Produced in small numbers.

167. A reworked Churchill II used as a trials vehicle for the Onion device.

168. Churchill AVRE with Goat equipment approaching obstacle to be breached. This is the later Mk III type Goat.

(3) Miscellaneous Types

Ardeer Aggie: This was an experimental prototype designed with a view to improving on the power of the Petard mortar fitted in the AVRE. The Ardeer projector was a recoilless gun in which recoil was neutralised by firing a dummy projectile rearwards simultaneous with the discharge of the main projectile. Design started in September 1943 and a prototype was converted from a Churchill III. The projectile weighed 54lb, range was 450 yards, and the projector was 10ft long. Trials showed the weapon to be impractical under combat conditions, and it was abandoned.

169. Churchill AVRE with elevatable Goat for placing explosive charges on high obstacles.

170. Churchill AVRE propelling Dalton Mobile Bridge with second vehicle acting as carrier.

171. Churchill II with Carrot explosive charge (light type).

172. Churchill with Bangalore Torpedoes.

Woodpecker: Experimental version of the AVRE with four Petard mortars mounted on each side of the hull. They could fire in salvo or in single shot.

Churchill Snake: This was developed in 1942 and consisted of explosive-filled 3in pipe, carried in eight lengths above each track. To clear a path through a minefield, the piping was joined together, pulled or pushed over the minefield and detonated, so clearing a path through the mines. The carrier vehicle was either a Churchill III or IV, or a Churchill 3in Gun Carrier (qv) with gun removed, as illustrated.

Churchill AVRE with Conger 2in, Mk I: The Conger was similar in principle to the Snake, consisting of 330 yards of 2in hose-pipe. It was carried in an engineless Universal Carrier towed behind the AVRE. The Carrier was towed to the edge of the minefield, the Conger was fired across by rocket, explosive liquid was pumped in to the hose from the Carrier, and the complete pipe was detonated.

Explosive Devices: After the Dieppe raid, various methods were devised whereby explosives could be carried up to sea walls and other fortifications, placed in position, and detonated from a distance. These all involved a form of framework on the front of a Churchill (though the Matilda was also used in trials) to hold the explosive. The Carrot simply allowed the charge to be dumped in position, the Onion allowed a charge to be hung from an obstacle, and the Goat was a more precise type of carrier with hinged frame which tilted on contact with the obstacle to ensure the charge slid into contact. The elevatable Goat was a similar equipment on a long angled carrier for reaching high obstacles like the top of sea walls. Only the Goats achieved production status and were carried on the AVRE. A further experimental device was the Churchill with Bangalore Torpedoes. This device consisted of two lengths of Snake piping fitted to an Onion frame assembly and was for use against light obstacles and barbed wire.

173. Churchill AVRE (far left) propelling Mobile Bailey Bridge.

174. One of the six Black Prince pilot models.

FIRST attempt to mount a large calibre high velocity gun in the Churchill chassis resulted in the Churchill 3in Gun Carrier, proposed and developed from September 1941. Production was limited by a reversal of policy in favour of concentrating on Churchills with 6pdr guns in 1942. For a variety of reasons, including limited numbers and indecision as to employment of a limited-traverse vehicle of this kind, the entire project was dropped in 1942 and the 3in Gun Carriers were later used to carry Snake mine detonating equipment. Meanwhile the A30 Challenger was developed as a cruiser tank with 17pdr gun in a widened, lengthened, Cromwell chassis. By September 1943, however, it had become clear that defects in the Challenger design were going to delay production at the crucial period when tanks with 17pdr guns were urgently needed for the forthcoming invasion of Europe. Design work on the new A41 (Centurion) had only just begun, and Vauxhall were therefore asked to go ahead and produce a version of the Churchill armed with the 17pdr gun as an interim vehicle with this armament. A proposal for doing this had been briefly considered and rejected two years previously in favour of the Gun Carrier, the main drawback being the narrowness of the Churchill hull which precluded the fitting of a turret wide enough to take the 17pdr.

Designated A43, the new vehicle was initially known as a "Super Churchill", but was later officially named Black Prince. While it utilised Churchill VII mechanical components as much as possible, the A43 involved much redesign work, mainly because of the wider hull required. The same Bedford twin-six engine was used as in the Churchill and this, combined with the increased weight (to 50 (long) tons), reduced the A43's top speed to only 11mph. Similar in external appearance to the Churchill VII, the Black Prince had wider (24in) tracks, and the air intakes were mounted on the hull top instead of at the side, with the engine exhausting at the hull rear. Six A43 prototypes were ordered and these were delivered for trials in May 1945, too late for combat as hostilities in Europe had ceased. Though given a full test programme, no production order was placed since the A41 Centurion was ready at the same time and proved a much superior vehicle.

SPECIFICATION

Designation: Tank, Infantry, Black Prince (A43)
Crew: 5 (commander, driver, gunner, loader, co-driver-hull gunner)
Battle weight: 112,000lb
Dimensions: Length 28ft 11in; Height 9ft; Width 11ft 3½in; Track width 24in; Track centres/tread —
Armament: Main: 1 × 17pdr OQF
Secondary: 2 × 7·92 cal Besa MG (one co-axial)
Armour thickness: Maximum 152mm
Minimum 25mm
Traverse: 360°. Elevation limits: —
Engine: Bedford twin-six 350hp
Maximum speed: 11mph
Maximum cross-country speed: 7mph (approx)
Suspension type: Sprung bogies
Road radius: 80 miles (approx)
Fording depth: 3ft 4in (unprepared)
Vertical obstacle: 2ft 6in
Trench crossing: 10ft
Ammunition stowage: —
Special features/remarks: Distinguished from Churchill VII by longer gun, bigger turret, wider hull, and absence of side-mounted air intakes. Though powerfully armed and armoured, this vehicle's tactical value was limited by its very slow speed.

175. TOG 2 as completed with 77mm gun.

TWO months prior to Britain's declaration of war against Germany in September 1939, the British Minister of Supply, responsible for armaments production, discussed likely tank requirements for any future conflict in Europe with Sir Albert Stern who had headed the British Tank Supply Dept in World War I. Resulting from this, on September 5, 1939, Sir Albert was asked to get together a committee of experts to study requirements and design problems. Among those invited to serve on the committee were Sir E. Tennyson d'Eyncourt, General Swinton, Mr Ricardo, and Major Wilson, all of whom had played a prominent part in tank development and production in World War I. The committee decided to request the General Staff to give an outline specification for a heavy tank, and Stern asked Sir William Tritton of Foster's of Lincoln, who had been responsible for much of the British tank output in 1916-18, to co-operate with the committee on any necessary research.

The General Staff suggested that members of the committee also visited France to look at the latest French tanks, and meet staff officers of the BEF. Meanwhile an outline specification was supplied and the committee was officially named, in October 1939, the "Special Vehicle Development Committee of the Ministry of Supply". The specification was similar to that issued to Harland & Wolff for the A20 "shelled area" infantry tank (see Churchill section). This called for a vehicle with all-round track able to cross shell-torn ground with armour proof against 47mm and 37mm anti-tank guns, and 105mm howitzers, at 100 yards range. It was to be armed with a field gun in the hull front to demolish fortifications, have sponsons mounting 2pdr guns, and have all arcs covered by Besa machine guns. It was to have a range of at least 50 miles, a speed of 5mph, and be diesel powered. Crew was to be 8 men. Finally it had to be transportable by rail.

Fosters drew up a design in December 1939 and a wooden mock-up was built and inspected. No suitable diesel engine was immediately available, however, so it was proposed to use a Paxman-Ricardo 450hp V12 diesel, developed to give 600hp. Due to the weight of the vehicle, electric transmission was suggested, at least in the prototype and English Electric Co were asked to supply a suitable type. The vehicle was known as TOG I (TOG: "The Old Gang", with reference to the committee members) and construction began in February 1940, the one and only prototype appearing in October 1940. It had a maximum speed of 8½mph and weighed about 50 tons without guns or sponsons. By this time the design had been modified, dispensing with the sponsons, but incorporating a turret for a 2pdr gun taken from an A12 Matilda tank. The gun in the nose was a French 75mm howitzer and mount as fitted in the Char B. Tracks were unsprung and almost identical to those used in the later British World War I tanks. Trials showed that the electric transmission system was unsatisfactory, and, in fact, the vehicle burnt out the motors on test. In TOG I, the diesel engine drove two generators, which in turn powered a motor for each track. The steering wheel operated a potentiometer which increased voltage on one or other of the motors to speed up the appropriate track to turn the vehicle. This was an ingenious but complicated system which, however, subjected the track and drive to great strain. TOG I was therefore re-built with hydraulic transmission which also proved unsatisfactory due to the time lag involved in filling the fluid couplings; this made steering hazardous. In its new form the vehicle was re-designated TOG IA.

While TOG I was being built, design of an improved model was put in hand which had recessed tracks in order to reduce the height of the hull. This was TOG 2, of which the one and only model appeared in March 1941. This was to have a larger turret with a 6pdr gun, plus the side sponsons as designed. However, the sponsons were never fitted, and the turret installed for initial trials was a mock-up with dummy gun, differing from the low turret envisaged in the working drawings for this vehicle. In fact, the turret appears to have been that intended for TOG 2R (R: revised), a proposed development which was about 6ft shorter than TOG

2, omitted the side sponsons entirely, and had torsion bar suspension. TOG 2R was never built, though a turret did materialise to replace the mock-up on TOG 2. This vehicle was mechanically similar to TOG 1.

TOG 2 was used for trials, but in the intervening period since the TOG was suggested, the A22 Churchill had appeared and been accepted and produced as the standard heavy infantry tank. Interest in the TOG therefore declined, but in early 1942 TOG 2 was fitted with a new turret and 17pdr gun for tests, the turret, designed by Stothert & Pitt of Bath being intended for the Challenger, then under development. TOG 2, now re-designated TOG 2* was thus the first British tank to mount a 17pdr gun. In modified form, the turret and its associated Metadyne traverse system was fitted in the Challenger.

SPECIFICATION

Designation: Tank, Heavy, TOG
Crew: 6–8 (TOG2*) (driver, commander, gunner, loaders (2), co-driver)
(TOG1) (driver, commander, gunner, loader, sponson gunners (4))
Battle weight: 179,200lb (TOG 2*)
142,320lb (TOG I)
Dimensions: Length 33ft 3in; Height 10ft; Width 10ft 3in; Track width —; Track centres/tread —
Armament: Main: 1 × 17pdr OQF (TOG2*)
1 × 6pdr (77mm) (TOG2)
Secondary: None actually fitted
Armour thickness: Maximum 50 plus 25mm
Minimum 25mm
Traverse: 360°. Elevation limits: —
Engine: Paxman-Ricardo V12 diesel 600hp (with electric transmission)
Maximum speed: 8·5mph
Maximum cross-country speed: 4mph (approx)
Suspension type: None–rigid rollers.
Road radius: 50 miles (approx)
Fording depth: —
Vertical obstacle: 7ft
Trench crossing: 12ft
Ammunition stowage: —
Special features/remarks: Vehicles built to an outmoded concept which nonetheless exhibited ingenious but complicated mechanical features. The very long ground contact in relation to width made steering very difficult.

176. TOG 2 with Stothert & Pitt turret and 17pdr gun.

177. TOG 1 with 75mm gun in nose and Matilda turret.

178. TOG 2 as first built with mock-up turret and 6pdr gun.

AS stated elsewhere in this book, the experience of the desert fighting of 1941-42, plus the appearance of US medium tanks in British service, led to a major revision of tank policy by the British General Staff in the fall of 1942 with the result that reliability and speed were now considered more important than heavy armour protection. Work was also initiated in developing a 75mm gun with "dual purpose" HE/AP capability, and it was for the first time that there was a need for a "universal" chassis capable of adaptation for the old "infantry", "cruiser" and other AFV roles. Hopes were thus pinned on the A27 series, then in an advanced state of development and about to enter production. The A27M (Cromwell) with its Meteor engine had proved particularly successful on trials and was to go into large scale production. Since the Churchill tank was both slow and mechanically unreliable at that time, it was proposed to cease production of this vehicle in 1943 once Cromwells were available in adequate numbers. Churchill production facilities would then also be switched to Cromwell manufacture. As an interim measure, therefore, while a new "universal" chassis design was contemplated, consideration was given to using the Cromwell chassis as a basis for a heavier vehicle for the "infantry tank" role to replace the Churchill. Rolls-Royce drew up schemes for two possible developments on these lines. The A31 was simply a Cromwell with extra armour added while the A32 was a more ambitious adaptation with armour basis brought up to A22 (Churchill) standard and a new, stronger, suspension to compensate for the extra 4½ (long) tons of weight this modification involved. Neither of these Rolls-Royce projects were followed up, however, but a third, by English Electric, designated A33, was produced as a pilot model. The English Electric design used a basic A27 hull and turret with added armour, married to the T1 track and suspension used on the American M6 heavy tank (qv). English Electric, incidentally, had participated in the original design work on the A27 series. The pilot model was completed in 1943, armed with a 6pdr gun though a 75mm gun was envisaged in the original specification. A second pilot model was also built with widened Cromwell-type tracks replacing the T1 tracks, and different side skirts. Designed by the LMS, this vehicle's suspension was known as "R/L Heavy" type. By this time, however, the Churchill had vindicated itself in the Tunisian and Italian campaigns and remained in production. Thus the A33 requirement lapsed and there was no production order.

SPECIFICATION

Designation: Tank, Heavy Assault, A33
Crew: 5 (commander, driver, gunner, loader, co-driver)
Battle weight: 100,800lb
Dimensions: Length 22ft 8in; Height 7ft 11in; Width 11ft 1½in
Track width —
Track centres/tread —
Armament: Main: 1 × 75mm OOF
Secondary: 2 × 7·92 cal Besa MG
Armour thickness: Maximum 114mm
Minimum 20mm
Traverse: 360°. Elevation limits: —
Engine: Rolls-Royce Meteor V12 600hp
Maximum speed: 24mph
Maximum cross-country speed: 12mph (approx)

180. A33 second pilot with British "R/L Heavy" type tracks.

Suspension type: See text.
Road radius: 130 miles (approx)
Fording depth: 3ft
Vertical obstacle: 3ft
Trench crossing: 7ft 6in
Ammunition stowage: 64 rounds, 75mm
Special features/remarks: Hull and turret basically the same as A27 series with added armour. Second pilot, with British tracks, had deeper side skirts concealing top run of track. Work on these vehicles was abandoned in May 1944.

INFANTRY TANK, VALIANT (A38) — United Kingdom

181. Valiant I pilot model in mild steel.

IN late 1943, Vickers suggested a design for an improved version of the Valentine which would incorporate as many existing Valentine components as possible and overcome the main failing of the Valentine design by having a larger three-man turret. Designated A38, this vehicle was to use an uprated GMC diesel engine plus other mechanical parts of late-production Valentines, be armed with a 6pdr gun (or later the 75mm gun), and be of all-welded construction. Hull shape and dimensions were very similar to the Valentine, but the hull front was made up of castings instead of flat plates. The turret was both wider and longer than the Valentine's, made of castings bolted together. Finally the suspension was modified to have all wheels the same size and independently sprung, thus simplifying maintenance and production. To leave Vickers' resources clear for other work, however, detail design parentage was passed to Birmingham Carriage & Wagon briefly, then on to Ruston & Hornsby who completed the pilot model in mid 1944. Named Valiant, the A38 was about 10 (long) tons heavier than the Valentine and thus slower. To overcome this a second pilot vehicle was put in hand, Valiant II, with a Rolls-Royce Metorite engine as the intended power unit. This was simply a shortened version of the Meteor engine with 8 instead of 12 cylinders to fit in the restricted space of the Valiant's engine compartment. However, with the closing stages of hostilities in Europe and no further requirements for infantry tanks of this type, the A38 scheme was dropped in 1945.

SPECIFICATION
Designation: Tank, Infantry, Valiant (A38)
Crew: 4 (commander, gunner, loader, driver)
Battle weight: 60,480lb
Dimensions: Length 17ft 7in; Height 7ft; Width 9ft 3in; Track width —; Track centres/tread —
Armament: Main: 1 × 75mm OQF
Secondary: 2 × 7·92 cal Besa MG
Armour thickness: Maximum 114mm
Minimum 10mm
Traverse: 360°. Elevation limits: —
Engine: GMC diesel 210hp
Maximum speed: 12mph
Maximum cross-country speed: 7mph (approx)
Suspension type: Independently sprung bogie for each wheel.
Road radius: 80 miles (approx)
Fording depth: 3ft
Vertical obstacle: 2ft 9in
Trench crossing: 7ft 6in
Ammunition stowage: —
Special features/remarks: Good adaptation of Valentine design, but much underpowered by 1945 standards. No production order; trials only.

HEAVY ASSAULT TANK, TORTOISE (A39) — United Kingdom

182. The Tortoise first pilot model; machine guns are not fitted in this view.

THE Tortoise fell completely outside the sequence of British tank policy and procurement in the years 1939-45, though it was obviously partly inspired by the original "shelled area" tank concept of 1939 which gave rise to the A20/A22 and TOG designs. The idea for this very heavily armoured tank with a super-heavy gun was initiated mainly by Mr Duncan Sandys, then a junior minister at the Ministry of Supply, towards the end of 1942, and may have been influenced at this period by events in the Western Desert fighting when German tanks and anti-tank guns were giving existing British tanks a battering. Nuffield Mechanisations & Aero were entrusted with the design which included a 3·7in 32pdr gun in a limited traverse mount, and armour proof against any known German anti-tank gun. Impetus to the work on this vehicle was given in late 1944 by the appearance of the German King Tiger and Jagd Tiger. The vehicle was then given the designation A39, the design was approved, six pilot models were ordered, and a production order was promised. Provisional target date for completion of the first vehicle was August 1945.

However, the six pilot models were not delivered for trials until 1946-47, by which time interest in these vehicles had ceased and no further work was done on them. The Tortoise was of extremely limited tactical value, being too slow, too large, and offering severe transportation problems. There were several unique features in the design, however, including the cast one-piece superstructure, the twin Besa MG cupola specifically for AA defence, and the suspension of 16 pairs of bogies each side.

183. Tortoise first pilot model; side view.

SPECIFICATION

Designation: Tank, Heavy Assault, Tortoise (A39)
Crew: 7 (commander, driver, co-driver, gunner, machine gunner, loaders (2))
Battle weight: 174,720lb
Power/weight ratio: 7·7hp/ton
Dimensions: Length 33ft (Including gun) Track width—
Hull length 23ft 9in
Height 10ft Track centres/tread—
Width 12ft 10in
Armament: Main: 1 × 32pdr OQF
Secondary: 3 × 7·92 cal Besa MG
Armour thickness: Maximum 225mm
Minimum 35mm
Traverse: 20° right, 20° left. Elevation limits: —
Engine: Rolls-Royce Meteor V12 600hp
Maximum speed: 12mph
Maximum cross-country speed: 4mph (approx)
Suspension type: Box bogie
Road radius: —
Fording depth: —
Vertical obstacle: —
Trench crossing: —
Ammunition stowage: —
Special features/remarks: Six pilot vehicles only, all completed post-war. Detail differences in individual vehicles included slight variations in superstructure shape and the addition of smoke dischargers in some. Driver sat in right superstructure front. Independent machine gun turret on roof with two Besa guns. Muzzle velocity of 32pdr gun was 3·050ft/second; largest gun fitted to any British AFV of 1939-45 design.

PART 2

AMERICAN VEHICLES

184. The first US tank unit to arrive in Britain in 1942, training with late production M3 medium tanks.

185. Standard production M1 Combat Car.

WITH the general run-down of American forces at the end of the First World War, the American General Staff disbanded the Tank Corps in 1919 and under the National Defense Act of 1920, tanks and tank development (in conjunction with the Ordnance Department) became an infantry responsibility. The General Staff subsequently defined the role of the tank in future war "to facilitate the uninterrupted advance of the rifleman in the attack". For economic and operational reasons, future tanks were to be concentrated on "light" and "medium" types, the former restricted to 5 tons in weight so as to be transportable by truck, and the latter limited to 15 tons to meet military bridging restrictions. Severe fiscal limitations imposed throughout the twenties allowed the production of only about two experimental tanks a year, however, culminating in the T1E4 light tank of 1931 which established the layout of rear engine and front sprocket drive adopted in all subsequent US light tanks. In 1927 the American General Staff had set up a very small experimental Mechanised Force, largely influenced by the similar unit which the British had just established, drawn mainly from infantry tank units and with the emphasis on light tanks. By 1931, however, General MacArthur had become US Army Chief of Staff and promoted mechanisation throughout the army. Among other things, MacArthur decided that in the mechanised age the US Cavalry, equipped with tanks and armoured cars, had an "exploiting" role in armoured warfare quite distinct from the infantry support role envisaged for tanks in the US Army until that time. The cavalry took over the Mechanised Force and was authorised to equip itself with tanks. However, to conform with the 1920 Defense Act cavalry tanks were known as "combat cars", a legal formality necessary to overcome the rule making tanks a prerogative of the infantry.

By 1934-35, three more experimental light tanks had been produced, T2, T2E1 and T2E2. The T2 itself had features inspired by the Vickers Armstrong 6 ton tank including the Vickers type leaf spring suspension.

For economy reasons, it was desirable that the light "combat cars" for the cavalry should be adapted from the infantry's light tanks. Concurrent with the T2 light tank, Rock Island Arsenal produced a similar vehicle, therefore, for cavalry use, the T5 combat car. This differered from the T2 principally in having vertical volute spring suspension instead of leaf spring suspension. Development led to the T5E2 and, under the designation M1 Combat Car, this vehicle entered service with the US Cavalry in 1937. Armament of the M1 was a ·30 and ·50 calibre machine gun in the turret and another ·30 calibre weapon in the hull front. Engine was a 7 cylinder Continental gasoline type which gave a top speed of 45mph, the crew was four men, and weight 9·7 (short) tons. An improved model, the M2 Combat Car, introduced a trailing idler to give a lengthened ground contact and improved ride.

In July 1940, the new Armored Force was created, abolishing the division between infantry and cavalry tank units. The "combat car" term was now no longer necessary and the M1 and M2 Combat Cars were accordingly redesignated Light Tank M1A1 and M1A2. These vehicles were not used operationally by American forces in World War II, but some were used for training at Fort Knox and other centres. The M1/M2 Combat Car series was, however, important as the basis for most subsequent US light tank designs until 1944, and much useful operating and design experience was gained with these early models.

VARIANTS

Combat Car M1A1: 1938 production model with constant mesh transmission (M1 had sliding gear type). 17 built.

Combat Car M2: As M1A1, but with improved turret and fitted with Guiberson T1020 diesel air-cooled radial engine in place of Continental gasoline type of M1 and M1A1. This vehicle was redesignated Light Tank M1A1 in 1940. 7 built. Trailing idler was introduced on this model.

Development Vehicles

M1E1 was prototype for series; M1E2 was prototype for M1A1; M1A1E1 was prototype for M2; M1E3 was experimental version of M1 (1939) tested with continuous band rubber tracks, and later with rubber-block tracks and lower gearing.

186. M2 Combat Car (later Light Tank M1A1).

Specification

Designation: Combat Car M1 (Light Tank M1 from July 1940)
Crew: 4 (commander, turret gunner, driver, hull gunner)
Battle weight: 19,644lb
Dimensions: Length 13ft 7in; Height 7ft 9in; Width 7ft 10in; Track width 11½in; Track centres/tread 6ft
Armament: Main: 1 × ·50 cal MG
1 × ·30 cal MG
Secondary: 1 × ·30 cal MG (hull)
Armour thickness: Maximum 16mm
Minimum 6mm
Traverse: 360°. Elevation limits: —
Engine: Continental W-670 gasoline, 7 cylinder 250hp
(Guiberson T1020 diesel in M2)
Maximum speed: 45mph
Maximum cross-country speed: 15–20mph (approx)
Suspension type: Vertical volute
Road radius: 100 miles
Fording depth: 4ft 4in
Vertical obstacle: —
Trench crossing: —
Ammunition stowage: 1,100 rounds ·50 cal MG
6,000 rounds ·30 cal MG
Special features/remarks: Forerunner (with M2 light tank) of subsequent American light tank designed until 1944. Obsolete by 1940, and replaced by late M2 series vehicles. Riveted construction.

187. M1 Combat Car of US 7th Cavalry at speed.

188. M2 Combat Car showing trailing idler.

189. A production M2A4, showing the three hull-mounted machine guns and the vision ports in the turret and cupola.

CIRCUMSTANCES leading to the development of the original T2 pilot model light tank are related in the M1 Combat Car section. Designed and built at Rock Island Arsenal, the T2 was produced in 1933. It had a simple riveted box-like hull with rear-mounted engine and drive to the front sprockets. These features were inherited from the later T1 series experimental tanks, but the suspension was copied from the Vickers 6 ton tank (qv) which had been demonstrated in America. Comparative trials with the contemporary T5 Combat Car (qv) showed, however, that the vertical volute suspension of this vehicle was much superior to the Vickers leaf spring suspension and the vertical volute suspension was fitted in a second light tank prototype, T2E1, produced after the trials in April 1934. For the infantry support role for which all American tanks were envisaged in the mid thirties, a machine gun armament was considered sufficient, and in the T2E1 this consisted of a ·30 cal and ·50 cal Browning mounted in a single turret which stretched nearly the whole width of the hull. A second ·30 cal machine gun was mounted in the hull front. The T2E1 was standardised late in 1935 and put into production at Rock Island Arsenal under the designation Light Tank M2A1, 19 vehicles being built in addition to the pilot model.

Meanwhile an improved vehicle, the T2E2 was built which was identical to the T2E1 but now had two turrets side by side, one for each machine gun, instead of the single full-width superstructure. This was standardised and put into production in 1936-37 as the Light Tank M2A2.

In 1938 improvements were incorporated in production vehicles to give better riding qualities, these consisting of longer stroke springs in the bogies with the rear bogies set 11 inches further aft. The improved model was designated Light Tank M2A3.

Light Tanks M2A1, M2A2, and M2A3 were all considered obsolete by 1940 and none was used in combat. They did, however, perform a useful training role in the early war years. Final versions of this type in production were the M2A2E3 and M2A3E3 which were vehicles re-engined with diesels.

Final and most important vehicle in the M2 light tank series was the M2A4. Comparative experience with the M2 light tanks and the corresponding M1 combat cars showed the advantages of the single turret in the latter. In 1939 Rock Island Arsenal designed an improved version of the M2A3 reflecting user experience. Designated M2A4, this was basically the M2A3 with a traversing turret replacing the twin turrets, a 37mm gun as main armament, three hull machine guns, and thicker armour with 25mm maximum. In September 1939 war was declared in Europe, giving impetus to the American rearmament programme. Rock Island Arsenal which had been responsible for the limited peace-time production of the US Army's tanks, guns, and munitions, lacked space and facilities for tank production on the huge scale now envisaged for future American tank requirements. The Ordnance Department had planned to contract with commercial heavy engineering firms if extensive tank production was required in emergency, and tenders were thus invited for building the new M2A4. The contract, initially for 329 vehicles, went to American Car & Foundry in October 1939 and the first production vehicle was delivered in April 1940. Meanwhile the order was increased to 365 vehicles and the final production M2A4 was eventually delivered in March 1941. The M2A4 was used in the early Pacific campaigns, but most were used for training.

190. Standard production M2A3 light tank; note the twin side-by-side turrets.

191. One of the M2A4 light tanks delivered to the British, shown with Browning ·30 cal machine gun in AA mount on turret.

British Service

A small batch of M2A4s were delivered to the British in 1941, but these were used only for training, vehicles of the M3 light tank series (qv) being issued to combat battalions.

Variants

M2A1: Initial production type with single fixed turret.
M2A2: Second production type with twin turrets. Machine guns in turret had 270° arc of fire, since the two turrets partly fouled each other.
M2A3: Third production type with improved suspension, slightly lengthened hull, thicker armour, improved engine access, increased gear ratios, better engine cooling, and many other minor detail changes. Twin turrets as in M2A2.
M2A4: Principal production type (see text). Hull as M2A3 but with additional ·30 cal MG in each side sponson, firing forward. Slightly uprated engine. Traversing turret with 37mm gun and co-axial ·30 cal MG replaced the two single turrets of the M2A3. Hand traverse. Vision ports in turret and cupola and mount for AA MG on turret rear. A few late production vehicles were fitted with Guiberson T1020 diesel engines in place of the Continental gasoline engine.

Development Vehicles

T2E1 was prototype for M2A1; T2E2 was prototype for M2A2; M2A2E1 was M2A2 test vehicle with Guiberson diesel engine installed; M2A2E2 was test vehicle with other detail changes introduced in M2A3; M2A2E3 was M2A2 with diesel engine and experimental trailing roller; M2A3E1 was M2A3 with Guiberson diesel installed; M2A3E2 was M2A3 test vehicle with experimental electric transmission; M2A3E3 was M2A3 re-engined with GM diesel motor.

Specification

Designation: Light Tank M2A1, M2A2, M2A3, M2A4.
Crew: 4 (commander, driver, co-driver, gunner)
Battle weight: 18,790lb (M2A1), 19,100lb (M2A2), 20,000lb (M2A3), 23,000lb (M2A4)
Dimensions: Length 13ft 7in (M2A1/2) 14ft 6in (M2A3/4) Track width $7\frac{5}{8}$in
Height: 7ft 4in (M2A1/2/3) 8ft 2in (M2A4)
Width: 8ft $1\frac{1}{4}$in Track centres/tread 6ft 1in
Armament: Main: See text for each variant
Secondary: See text for each variant
Armour thickness: Maximum 16mm (25mm on M2A4/5) Minimum 6mm
Traverse: 360° (M2A4/5) 270° arc (others)
Elevation limits: +20° to —10°
Engine: Continental W-670 gasoline, 7 cylinder, 250HP (Guiberson T1020 diesel in some vehicles as noted; GM diesel in M2A2E3)
Maximum speed: 25-30mph
Maximum cross-country speed: 18mph (approx)
Suspension type: Vertical volute
Road radius: 130 miles
Fording depth: 3ft 8in
Vertical obstacle: 2ft
Trench crossing: 6ft
Ammunition stowage: 103 rounds, 37mm (M2A4) 2,137 rounds ·50 cal MG 7,185 rounds ·30 cal MG (M2A3)
Special features/remarks: A few of the diesel engined M2A2E3 vehicles are believed to have been used in Burma. All others, except M2A4 used for training only. M2A4 distinguished from the remainder by its turret and 37mm gun. All were of riveted construction.

192. Standard production M3 light tank, showing all-riveted construction.

THE Light Tank M3 was a progressive improvement on the M2A4 designed at Rock Island Arsenal in the spring of 1940 and incorporating lessons observed from the tank fighting in Europe in the 1939-40 campaigns. The main requirement was for increased armour thickness, but this in turn called for stronger suspension. Maximum frontal armour was increased to 38mm (51mm on the nose) and the vision ports in the turret sides were eliminated. A large trailing idler was fitted to increase ground contact. Other modifications included a lengthened rear superstructure and improved armour on the engine covers as a precaution against strafing from the air. The M3 was approved and standardised in July 1940 and entered production in March 1941 at American Car & Foundry, being introduced straight on to the line after completion of the M2A4 contract.

Several further improvements were made during production, first of these being a welded turret, replacing the riveted type, which was developed in late 1940 and introduced almost immediately into production vehicles in March 1941. This change was mainly to reduce weight, though it also eliminated the danger of "popping" rivets in event of a hit. A further change was introduced in early 1941 with a welded/cast homogenous turret of rounded shape replacing the multi-faced turret used until then. This was later incorporated into production vehicles. From mid 1941 a gyro-stabiliser was fitted for the gun and in the fall of 1941, following British experience with M3 light tanks in the North African desert fighting, two 25 gallon jettisonable fuel tanks were introduced to increase the range. From early 1942 an all-welded hull was adopted. To ease engine supply problems, 500 M3 light tanks were completed with Guiberson T1020 diesel engines replacing the standard Continental petrol engine. These vehicles were sometimes called M3 (Diesel). Externally they were identical to the standard M3.

A further improved model was designed, tested, approved and standardised in August 1941. This had the cupola eliminated to reduce overall height, a gyro-stabiliser for the gun, power traverse for the turret, and a turret basket. Designated M3A1 it was introduced to the American Car & Foundry production line in June 1942 to follow on the M3 which finally went out of production in August 1942.

A further change in the M3A1 was the elimination of the two sponson machine guns carried in the M2A4 and M3. These were fired remotely by the driver and proved of limited value, being finally sacrificed to reduce weight, and increase interior stowage. The British had already removed these guns from many of the M3s delivered to them.

Final production variant was the M3A3 which represented a radical redesign with a new all-welded hull enlarged by extending the side sponsons and the driver's compartment forward and upward. This gave room for extra fuel tanks and increased ammunition stowage. Sandshields (another lesson from the desert fighting) were added and numerous other detail changes were made. Standardised in August 1942, the M3A3 entered production in early 1943.

193. M3 light tank with welded turret shown on delivery to Britain (British Stuart I).

SPECIFICATION
Designation: Light Tank M3, M3A1, or M3A3
Crew: 4 (commander, gunner, driver, co-driver)
Battle weight: 27,400lb (M3), 28,500lb (M3A1), 31,752lb (M3A3)
Dimensions: Length 14ft 10¾in (16½ft in M3A3) Track width 11⅝in Track centres/tread 6ft 1in
Height 8ft 3in (7ft 6½in in M3A1 and M3A3)
Width 7ft 4in (8ft 3in in M3A3)
Armament: Main: 1 × 37mm M5 or M6 gun
Secondary: 3 × ·30 cal Browning MG (plus 2 others in sponsons in M3)
Armour thickness: Maximum 51mm
Minimum 10mm
Traverse: 360°. Elevation limits: +20° to −10°
Engine: Continental W-670, 7 cylinders, petrol or Guiberson T1020, 9 cylinders, radial, diesel
Maximum speed: 36mph
Maximum cross-country speed: 20mph (approx)
Suspension type: Vertical volute and trailing idler
Road radius: 70 miles
Fording depth: 3ft
Vertical obstacle: 2ft
Trench crossing: 6ft
Ammunition stowage: 103 rounds (M3), 116 rounds (M3A1), 174 rounds (M3A3), 37mm
6,400–8,270 rounds ·30 cal MG
Special features/remarks: Most widely used American-built light tank type. Fast and reliable vehicle for the "recce" role for which it was intended, but did not lend itself to special purpose adaptation, being too light and underpowered for the weights involved. Declared obsolete in US Army in July 1943, but remained in service until the War's end (and beyond) by other users. Important addition to British tank strength in the Western Desert fighting of 1941-42

VARIANTS

M3: Initial production type (see main text) with riveted turret, welded turret, cast/welded turret, and welded hull introduced successively during production run. Total production: 5,811. Continental petrol or Guiberson diesel (500 vehicles only).

M3A1: As late-production M3 but with gyro-stabiliser, turret basket, and no cupola (see main text). Total production 4,621. Continental petrol or Guiberson diesel (211 vehicles only). Pilot and early production models had riveted hulls; remainder were all-welded. Sponson machine guns eliminated.

M3A2: Projected designation, March 1942, for the all-welded version of the M3A1. In the event no distinction was made between riveted and welded vehicles.

M3A3: Final production type with redesigned hull, all-welded, with increased internal stowage (see main text). Total production 3,427.

M3 Command Tank: Field adaptation (any variant) entailing removal of turret and substitution of welded box superstructure with ·5 cal Browning MG on flexible mount. For appearance see M5 Command Tank. Used by senior officers.

M3 with Maxson Turret: 1942 project to mount Maxson quad ·5 cal MG turret on an M3 light tank in place of the standard turret to provide an AA tank. Tested but rejected in favour of similar equipment mounted in a half-track.

M3 and T2 Light Mine Exploder: 1942 experiment with T2 mine exploder on boom rigged in front of M3 light tank. Too unwieldy for vehicle and thus rejected.

194. M3 with cast/welded homogenous rounded turret and riveted hull.

195. M3 with cast/welded homogenous turret and all-welded hull. Machine guns not mounted.

196. M3A1 pilot model showing cupola eliminated and riveted hull. Later vehicles had all-welded hull as plate 195.

M3 or M3A1 with Satan Flame-gun: Flame-thrower vehicle produced in Hawai for use by USMC in Pacific Theater for attacking Japanese bunkers. The 37mm gun was removed from obsolescent M3 light tanks and replaced with flame projector and Canadian Ronson flame-thrower equipment. Fuel capacity was 170 gallons and range 40-60 yards. A total of 20 were converted in late 1943.

M3A1 with E5R2-M3 Flame-gun: Portable flame-thrower modified to fit ball machine gun mount in hull front. Fuel reservoir stowed inside vehicle. Capacity was 10 gallons and the equipment was supplied for quick fitting in the field in place of the hull machine gun. This could also be used in the M5 light tank (qv).

197. Standard production M3A3 light tank, showing larger, re-designed, welded hull.

198. M3A3 light tank in British service (Stuart V) showing wading equipment fitted. Note extension at turret rear for radio equipment another feature introduced in the M3A3.

T18 75mm Howitzer Motor Carriage: Design initiated in September 1941 to provide a close-support vehicle utilising the M3 light tank chassis. It consisted of a 75mm M1A1 pack howitzer in a mount adapted from that used in the M3 medium tank, fitted in the right front of the vehicle with built-up superstructure. Two test vehicles were produced with mild steel superstructure and sent for trials to Aberdeen Proving Ground. They proved unsatisfactory due to the high superstructure and nose heaviness. Project was abandoned in April 1942 in favour of the M8 Howitzer Motor Carriage (qv).

T56 3in Gun Motor Carriage: Project initiated in September 1942 to mount 3in gun on modified M3A3 chassis. Engine moved to centre of vehicle and gun mounted at rear with shield for crew. Gun was, however, too heavy for the vehicle with adverse effect on performance, and there was only limited crew protection.

T57 3in Gun Motor Carriage: As T56 but with up-rated Continental engine from M3 medium tank and shield omitted in attempt to overcome the performance drawbacks of T56. Also proved unsatisfactory at APG tests and both T56 and T57 projects were dropped in February 1943.

199. The experimental M3 with Maxson quad ·5 cal turret.

Development Vehicles

M3E1 was M3 fitted experimentally with Cummins HBS diesel motor. M3E2 was test vehicle for twin Cadillac engines and Hydromatic transmissions installed in M5 light tank (qv), 1941. M3E3 was same vehicle with turret basket fitted to test installation for M3A3 and M5, 1942. M3E4 was test vehicle for trials with British Straussler DD equipment, 1942, as fitted to Tetrarch light tank in Britain. M3A3E1 was trials vehicle for Spicer Automatic Torque Converter Transmission. M3A3E2 was further trials vehicle with this transmission but with Continental R950 fitted to provide more power.

200. M3 with T2 Mine exploder rigged.

201. M3A1 light tank with Satan Flame-gun, used by USMC.

202. T18 75mm GMC, one of two prototypes produced.

203. T56 3in GMC, showing engine air vents moved to hull centre.

204. T57 3in GMC with crew platform lowered.

205. Stuart V converted to Kangaroo or Recce with folding canvas weather cover.

BRITISH SERVICE

The M3 light tank was the first American-built tank to see combat with British troops in World War II, 84 being sent to 8th Army in the first Lend-Lease shipment in July 1941. By November 1941, for the Operation Crusader battle, 163 were in service. Further deliveries of all basic models were made and M3 series light tanks were used by the British in Burma, Britain, NW Europe (from 1944), Italy, and North Africa. Several other countries (including Russia, China, and France) had M3 deliveries. In British Service, the M3 was named "General Stuart", more commonly called the "Stuart" and unofficially nick-named the "Honey". British designations and variants were as follows:

Stuart I: Basic M3 vehicle with Continental engine.

Stuart II: Basic M3 vehicle with Guiberson diesel engine. Also known as "Stuart Hybrid".

Stuart III: Basic M3A1 with Continental engine.

Stuart IV: Basic M3A1 with Guiberson diesel engine. Also called "Stuart Hybrid".

Stuart V: Basic M3A3 vehicle.

In British service these were modified by addition of British pattern smoke dischargers on turret sides, addition of sand shields in 8th Army service, and removal of sponson machine guns in most M3s.

Stuart Kangaroo: Late war conversion of redundant vehicles (any mark) by removal of turret and addition of seats for infantry for APC role in infantry units of armoured brigades. 1943-45 and post-war.

Stuart Recce: As Kangaroo but with various combinations of machine gun armament on extemporised pintle mounts. Turret removed.

Stuart Command: As Kangaroo but with extra radio equipment for unit commander.

Various forms of extemporised grenade netting could be seen fitted in some cases to these turretless conversions.

Stuart 18pdr SP: Middle East forces improvisation on at least one vehicle, with obsolete 18pdr field gun, less wheels, mounted in place of turret.

206. Stuart Command tank (on Stuart I chassis) of Grenadier Guards, Tunis 1943, carrying General Montgomery, 8th Army commander.

LIGHT TANK, M3

207. Stuart Recce (on Stuart I chassis) in Burma, 1944, with improvised grenade netting.

208. Fitting improvised mounts for ·30 cal MGs to Stuart Recce tank (on Stuart V chassis) in Normandy, June 1944. Note also improvised shield.

209. Stuart I 18pdr SP. Note gun carriage axles in improvised cradle.

210. Top detail view of M3 light tank shows extra fuel tanks retrospectively fitted to increase range.

211. M3 (Stuart I) of a Canadian armoured regiment in Britain, 1943.

212. Top view of M3A3 clearly shows modified superstructure shape of this model.

213. M3 light tank in service as British Stuart I showing modifications made by British, including smoke dischargers and sand shields.

LIGHT TANK, M5 SERIES, GENERAL STUART

United States

214. Standard production Light Tank M5A1.

LIGHT Tank M5 stemmed from a suggestion by Cadillac Division of GMC to the Ordnance Department that they should try the M3 light tank with twin Cadillac engines installed and the commercial Cadillac Hydra-matic transmission which was produced for automobiles. In the fall of 1941, a standard M3 was converted as a trials vehicle (the M3E2) to test the idea and proved most successful. This made a trouble-free 500 mile trial run, and the Cadillac-powered vehicle proved easy to drive, and smooth to operate. Due to the always acute shortage of Continental engines, the Cadillac modified vehicle was approved for production and standardised as the Light Tank M5 in February 1942. It was originally to be designated Light Tank M4, but this was changed to M5 to avoid confusion with the M4 medium tank (Sherman), then going into production.

To accommodate the twin Cadillac engines, the rear engine covers were stepped up. But the hull was otherwise similar in shape to that of the welded M3A1 apart from a sloping glacis. Turret installation, with basket and gyro-stabiliser, was tested in another development vehicle, M3E3, while in the following July another Cadillac facility, at Southgate, California, also commenced production. At the same time Massey-Harris commenced M5 production in Racine, Wis, under the "parentage" of Cadillac, and finally, in October 1943, when M3 series production ceased, American Car & Foundry also switched to turning out M5s.

The M5A1 was designed and standardised in September 1942 to bring the M5 up to the standard of the much-improved M3A3. Among changes common with the M3A3 were a new turret with bulge at the rear for a radio installation, larger access hatches for the driver and co-driver, improved mount for the 37mm gun, and improved vision devices. In addition there was better water-sealing on the hatches, an escape hatch added in the hull floor, and dual traverse allowing the commander to train the turret while firing his AA machine gun. There were also detachable sand shields. A later modification was the provision of a detachable shield/fairing on the turret side to protect the AA machine gun mount. From early 1943, the M5A1 replaced the M5 on production lines.

SPECIFICATION

Designation: Light Tank M5 or M5A1
Crew: 4 (commander, gunner, driver, co-driver)
Battle weight: 33,000lb (M5), 33,907lb (M5A1)
Dimensions: Length 14ft 2$\frac{3}{4}$in Track width 11$\frac{5}{8}$in
(15ft 10$\frac{1}{2}$in over stowage box in M5A1)
Height 7ft 6$\frac{1}{2}$in Track centres/tread 6ft 1$\frac{1}{2}$in
Width 7ft 4$\frac{1}{4}$in
(excluding sand shields on M5A1)
Armament: Main: 1 × 37mm gun M6
Secondary: 2 × ·30 cal Browning MG
(Plus AA MG in most vehicles)
Armour thickness: Maximum 67mm
Minimum 12mm
Traverse: 360°. Elevation limits: +20° to −10°
Engine: Cadillac Twin V8 220hp
Maximum speed: 36mph
Maximum cross-country speed: 24mph (approx)
Suspension type: Vertical volute
Road radius: 100 miles
Fording depth: 3ft
Vertical obstacle: 1ft 6in
Trench crossing: 5ft 4in
Ammunition stowage: 133 rounds (M5) and 147 rounds (M5A1) 37mm
6,250 rounds (M5) and 6,500 rounds (M5A1) ·30 cal MG
Special features/remarks: Most easily distinguished from M3 series vehicles by stepped up rear deck. Superior vehicle to M3 series, but not produced in such large numbers due to the appearance of the heavier M24 series. M5 series light tanks were re-classified "substitute standard" in July 1944 in the US Army.

LIGHT TANK, M5

215. M3E3 which was the prototype for the M5 series, a converted M3 light tank.

216. Late production M5A1 with sand shields and armour protection for AA mount.

VARIANTS

M5: Initial production type (see main text).

M5A1: Improved production type in line with improvements introduced in M3A3 (see main text).

M5 Command Tank: Vehicle with turret removed and replaced with box-like superstructure and ·50 cal Browning machine gun in flexible mount. For use by senior officers.

M5A1 with Psy-war equipment: Standard vehicle fitted with loud hailer and associated public address equipment for Psychological Warfare units, 1944-45.

M5 or M5A1 with Cullin Hedgerow Device: Standard vehicles with "prongs" made from beach obstacles by field maintenance units for cutting through "bocage" hedges and undergrowth, Normandy, June 1944.

M5A1 with E7-7 Flame-gun: Flame-thrower equipment replacing main gun. Short projector and fuel stowed in hull. Same equipment could be fitted in M3A1.

M5A1 with E9-9 Flame-throwing equipment: This was based on the British Crocodile idea (qv) and consisted of a flame projector replacing the hull machine gun, and fuel led, via an armoured pipe, from a towed trailer. Prototype only, started April 1943.

M5A1 with E8 Flame-gun: Developed from January 1943, this vehicle had the original turret removed, a new box-like superstructure added with a flame-gun carried in a small rotating turret at the top. It was a prototype only.

217. M5 light tank with Psy-war equipment, 1944.

218. M5 Command Tank, personal vehicle of commander, US 6th Armored Division, 1945. Note new drivers' hatches in glacis plate.

219. M5 with Cullin Hedgerow Device, Normandy, 1944.

M5 with T39 Rocket Launcher: T39 Rocket Launcher mounted on turret top and controlled in elevation and traverse by vehicle's gun and main armament. Fired 20 × 7·2 in rockets. Project only, 1944.

220. M5A1 with E7-7 Flame-gun.

221. The M5A1 with E9-9 Flame-throwing equipment, the sole prototype vehicle.

M5 Dozer: M5 light tank with turret removed and fitted with dozer blade at front end, 1944. A few vehicles so fitted retained the turret.

T27, T27E1 81mm Mortar Motor Carriage: This resulted from an army requirement for a mortar carrier on the M5A1 chassis. A prototype, T27, was built with turret removed and an armour superstructure added, 18in high and 25mm thick. Mortar was mounted to fire forward with 35 degree traverse and a ·50 cal MG was also fitted. An alternative design, the T27E1, was similar except that the 81mm mortar was mounted lower in the hull so that it did not project above the superstructure. Both were cancelled in April 1944, after tests, due to inadequate crew and stowage space.

T29 4·2in Mortar Motor Carriage: Following cancellation of the T27 vehicles, this was produced with the mortar space enlarged and utilising a more compact mortar. Crew space for operating the mortar was still found inadequate and the project was terminated in this form.

T8 Reconnaisance Vehicle: This was a conversion of redundant M5 light tank chassis in 1944 by removal of the turret and the fitting of a gun ring mounting a ·50 cal Browning MG in its place. Prototype for the conversion was known as the Reconnaisance Vehicle "B". T8s were in service as a "limited standard" type in 1944-45. A modified type with extra stowage racks added and rack for carrying land mines was designated T8E1.

DEVELOPMENT VEHICLES: M3E2 was M3 converted to test twin Cadillac engine installation. M3E3 was fuller conversion with addition of turret basket and other features for M5 and was M5 light tank prototype. M5A1E1 was an experimental vehicle with automatic 37mm gun and wider tracks which did not progress past trials status, 1943, due to development of later and better light tank designs. An

222. T39 Rocket Launcher installed on M5 light tank turret.

223. T27 Mortar Motor Carriage. T27E1 similar except for height of mortar mount.

224. T29 Mortar Motor Carriage.

225. T8E1 Reconnaisance Vehicle; T8 similar.

LIGHT TANK, M5

experimental AA tank version of the M5A1 was also produced with a twin ·30 cal Browning MG mount replacing the turret; it did not proceed past trials status. Finally M5s and M5A1s were also tested with various forms of flotation gear for amphibious operations, but none past the trials stage. For self-propelled vehicles based on the M5/M5A1 series, see separate entry.

BRITISH SERVICE

A small number of M5s and M5A1s were delivered to the British in 1943-44 and these were used in NW Europe from 1944.

Stuart VI: British designation for both M5 and M5A1 light tanks.

Some Stuart VIs were converted to Command, Recce, or Kangaroo versions by removal of turret. Details as for variants based on the M3 light tank (qv).

226. Stuart VI was British designation for M5 and M5A1, the latter shown here.

HOWITZER MOTOR CARRIAGE, M8

United States

227. M8 HMC in action in Belgium, late 1944.

REQUESTS from the Armored Force for a howitzer-equipped vehicle for the close support role, were met early in 1942 by the fitting of such a weapon to the M3 half-track as a temporary measure until a suitable mount could be developed on a full-tracked chassis. This early vehicle was designated T30 HMC, and was an adaptation with the standard M1A1 75mm pack howitzer. Aberdeen Proving Ground meanwhile adapted one of the first M5 light tank chassis to become available, fitting it with a 75mm howitzer on the centre line, mounted in the hull front. This was designated T41 Howitzer Motor Carriage and existed purely with a mock-up superstructure, unarmoured, for study. It was considered that this design gave inadequate protection to the crew and the T41 was abandoned in favour of a new design, the T47, which was also based on the M5 chassis.

The T47 was designed to give all-round traverse for the howitzer and superior crew protection. Basically it was an M5 light tank with the turret removed and replaced with a larger open-topped turret carrying the 75mm howitzer. A mock-up was produced in April 1942, approved, standardised as the M8 Howitzer Motor Carriage, and ordered into production. Cadillac built 1,778 production vehicles between September 1942 and January 1944. The M8 equipped the HQ companies of medium tank battalions until gradually replaced by the M4 medium tank with 105mm howitzer from the spring of 1944. It was used in NW Europe and Italy. Due to the turret size, drivers' hatches were provided on the hull front.

Details as for M5 light tank except for weight (36,000lb), elevation (+40° to −20°), and ammunition stowage (46 rounds, 75mm).

T82 Howitzer Motor Carriage: This was another SP vehicle on the M5A1 chassis, development of which began in December 1943 to provide a light weight SP howitzer for jungle warfare. To keep weight down, the turret was eliminated and the 75mm howitzer was fitted in a limited traverse mount in the hull front. Elevation was +30° to −5° and there was a limited traverse of a few degrees each side. Two pilot models were built for trials, but the project was abandoned in May 1945, when the requirement for such a vehicle lapsed. Details as for M5A1 except for armament and overall height.

228. One of the two T82 Howitzer Motor Carriage pilot models.

LIGHT TANK, T7 SERIES — United States

229. Light Tank T7E2 as first completed with 57mm gun.

IN the fall of 1940, consideration was given to a more powerful light tank to replace the M2A4 and M3 designs (qv), neither of which were considered entirely satisfactory by the new Armored Force. In January 1941 definite requirements, calling for a 14 (short) ton vehicle with 37mm gun, low silhouette, and 38mm armour maximum, were passed to the Ordnance Department and the construction of two pilot models was initiated at Rock Island Arsenal who also worked out the design. Designated T7, the first prototype was to have a welded hull, cast turret, modified vertical volute suspension, and 15⅛in wide tracks. For comparative purposes, the second pilot model, designated T7E1 was to be of riveted construction with a cast/welded homogenous turret, and horizontal (instead of vertical) volute suspension. This vehicle was, in fact, never completed, since riveted construction became obsolete during its development period. The T7E1 chassis was subsequently used for transmission and suspension trials. The same

LIGHT TANK, T7

Continental engine was to be used as in the M2A4 and M3.
A wooden pre-production mock up of the T7 was built at Rock Island, leading to a request from the Ordnance Dept for three further prototypes. These were T7E2 with cast hull top, cast turret, and Wright R-975 engine; T7E3 with welded hull and turret, automatic transmission, and twin Hercules diesel engines, and T7E4 with welded hull and turret and Cadillac twin engines and Hydra-matic transmission which Cadillac had by this time (September 1941) suggested for the M3 series. Detail changes requested increased the length and would bring the weight to 16 (short) tons. Of these, the T7E2 offered the best possibilities, and with a few further changes, the design was approved in December 1941. While the T7E2 pilot was being built it was decided to fit it with more powerful armament—the 57mm T2 (6pdr) gun—an adaptation of the British 6pdr—which was being fitted to the Canadian-built Ram tank(qv). A Ram turret ring was therefore incorporated and the vehicle was completed in June 1942 with a 57mm gun. At the same time, the Armored Force asked if the vehicle could be modified to take a 75mm M3 gun and the turret was duly re-designed to take this weapon. During development, armour thickness had been increased to a maximum of 63mm, and the other changes had brought the weight up to 25 (short) tons.

This took the vehicle out of the light tank class, and the Armored Force suggested it should be re-classified as a medium tank and standardised as the Medium Tank M7. This was done in late August 1942 and International Harvester Co were awarded a production contract for 3,000 vehicles to commence in December 1942. Meanwhile the pilot model in its final form was delivered to the Armored Force HQ, Fort Knox, for testing in early December and proved most unsatisfactory since its weight—almost double what was first envisaged—made it grossly underpowered. Fully stowed and with crew aboard it weighed even more, 29 (short) tons. In an attempt to provide more power, it was proposed to re-engine the vehicle with a Ford V-8 motor, the development project being designated M7E1.
By this time the M4 was in full production as the standard US medium tank and the Ordnance Department pointed out to the Armored Force that two distinct medium tank types, the M4 and M7, were unnecessary and would lead to complicated duplication of maintenance and production effort. Since the M7 had failed to come up to expectations it was agreed to drop this type and production was cancelled in February 1943 after only seven M7s had been completed. Work on the M7E1 was subsequently abandoned the following July and the T7/M7 series was declared obsolete at the end of 1943. The type was never used by the US Army.

LIGHT TANK, (AIRBORNE) M22, LOCUST

United States

230. T9E1 pilot model showing modified hull shape.

REQUIREMENT for an airborne tank was formulated at an Ordnance Department meeting in February 1941, attended by representatives of the Armored Force and USAAF, and an outline specification for such a vehicle was finalised in May 1941. It called for a tank of about 8 (short) tons in weight (half the weight of the M5A1 light tank), and correspondingly compact dimensions suitable for carriage either inside or beneath a transport aircraft. Design studies were invited from J. Walter Christie, GMC, and Marmon-Herrington. Of these, that submitted by Marmon-Herrington, was most promising and a pilot model, designated T9, was ordered. This design featured a 37mm M6 gun, a

Lycoming engine, an armour maximum of 25mm, and modified vertical volute suspension.

The T9 pilot model was delivered in the autumn of 1941 and in the light of trials, modifications were suggested including reshaping the hull front to improve shot deflection and the elimination of non-essential fittings, like the power traverse and gyro-stabiliser, to reduce overall weight. Two further pilot models, designated T9E1, were built incorporating these improvements and including an easily removable turret to facilitate air transportation. The modified design was ordered into production and 830 were built by Marmon-Herrington between March 1943 and February 1944. In September 1944 the T9E1 was redesignated Light Tank (Airborne) M22 and classed as "limited standard". The M22 was never used in combat by American forces, mainly because they lacked a suitable glider or transport aircraft to carry it. The only means evolved for air transportation of this vehicle by US forces was slung beneath the belly of a C-54 Skymaster transport plane. To do this the turret had to be removed and carried in the aircraft, being re-assembled with the tank on landing, which was a severe limitation to its tactical value.

231. Production M22 in British service, shown fitted with Littlejohn adaptor on gun barrel.

232. The original T9 pilot model.

British Service

The second T9E1 pilot model was shipped to Britain for airborne evaluation early in 1943. The British were designing the Hamilcar glider to carry the Tetrarch light tank (qv) in the airborne role and this also proved capable of carrying the M22, later named the Locust. A large number of M22s were supplied to the British under Lend-Lease for airborne operations and a handful of these were landed by Hamilcar glider in the Rhine crossing operation by the British 6th Airborne Division, March 24, 1945.

233. Production M22 showing method of attachment to belly of C–54 cargo plane for air transportation, turret removed.

Specification

Designation: Light Tank (Airborne) M22
Crew: 3 (commander, gunner, driver)
Battle weight: 16,400lb
Dimensions: Length 12ft 11in — Track width 11¼in
Height 6ft 1in — Track centres/tread 5ft 10½in
Width 7ft 1in
Armament: Main: 1 × 37mm M6 gun
Secondary: 1 × ·30 cal Browning MG
Armour thickness: Maximum 25mm
Minimum 9mm
Traverse: 360°. Elevation limits: +30° to −10°
Engine: Lycoming 0–435T 6 cylinder petrol 162hp
Maximum speed: 40mph
Maximum cross-country speed: 30mph
Suspension type: Vertical volute
Road radius: 135 miles
Fording depth: 3ft 2in
Vertical obstacle: 1ft ½in
Trench crossing: 5ft 5in
Ammunition stowage: 50 rounds, 37mm
2,500 rounds ·30 cal MG
Special features/remarks: Hull of rolled plate and cast turret. Four brackets on hull for slinging to belly of aircraft. These were removed on some vehicles in British service. Some British vehicles also had Littlejohn adaptors fitted to the 37mm gun (see illustration) to improve velocity. First 26 production vehicles had square driver's head cover, remainder had sloped sides. M22 was of conventional, but compact, design, but too thinly armoured and of limited tactical application.

234. M22 in British service leaving the hold of a Hamilcar glider during training, 1944. Note sand shields and the British type smoke discharger on turret side.

235. Production Light Tank T16.

MARMON-HERRINGTON produced a light weight two-man tank in the 4 ton class specifically as a Lend-Lease vehicle for delivery to China and the Netherlands East Indies. It had a crew of two (Commander/Driver) and an armament of a single ·30 cal Browning MG plus a second ·30 cal weapon for AA use. Of riveted construction, it had vertical volute suspension. Production totalled 240 vehicles, and the type was in limited use by the US Army. The designation T16 was allocated. Production was started in July 1942.

OTHER US LIGHT TANK PROJECTS

Light Tank T13: This was a vehicle designed by Allis-Chalmers and offered to the Ordnance Department in April 1942 as a projected light weight two-man vehicle weighing 3½ (short) tons and armed with a single 20mm Hispano-Suiza cannon. It was similar in size and layout to the T16. However, no requirement existed, or could be foreseen, in the US Army for a small vehicle of this type, and the idea was rejected, work on the design being subsequently stopped.

Light Tank (Amphibian) T10: This existed as a design study only, originating in April 1940 when a wheeled amphibious vehicle was proposed by the Quartermaster Corps. In 1941, the Ordnance Department put forward specifications, based on the original project, for an armoured version with turret and 37mm gun, 18mm armour maximum, and a weight of 12½ (short) tons. In May 1942 the design work was completed, but the vehicle, designated T10, was cancelled before a pilot model was built since no future requirement for this particular vehicle could be envisaged.

Light Tank T21: The Armored Force and Ordnance Department decided, in August 1942, to develop a light tank version of the Medium Tank T20 (qv), then being developed as a successor for the M4 medium tank series. The T21 specification, as drawn up in February 1943, envisaged a 24 (short) ton vehicle utilising the suspension of the T7 series (qv) or torsion bar suspension, armed with a 76mm gun, with armour maximum of 30mm, and with a hull and turret similar to the T20 medium tank. Two pilot models were projected, but authority to proceed with their construction was never given and the design was held in abeyance, mainly because the vehicle was too heavy for the light tank class.

236. Standard production M24 Chaffee.

OBSERVATIONS of the British experiences in the Western Desert fighting in 1942 when the 8th Army was using M3 series light tanks, showed that a heavier weapon was desirable for future US light tanks. A 75mm gun was fitted experimentally to a M8 HMC in place of the howitzer, and firing trials proved that it would be possible to develop a version of the M5 series light tank armed with the 75mm gun. Stowage space was severely restricted in the M5, however, more so with the fitting of a 75mm gun, and in addition the overall design of this vehicle was now dated and the armour thickness was inadequate. In April 1943, therefore, following the demise of T7 light/M7 medium programme (qv) the Ordnance Department, in conjunction with Cadillac (makers of the M5 series), began work on an entirely new light tank design which was to incorporate the best combinations of features from earlier designs with all lessons learned from previous experience. The twin Cadillac engines and Hydra-matic transmission which had been so successful and trouble-free in the M5 series were retained and the good accessibility which had been a feature of the T7 layout was adopted. A weight of 18 (short) tons was envisaged with an armour basis of only 25mm to save weight, but with all hull faces angled for optimum protection. Maximum turret armour was 37mm. Vertical volute suspension was replaced by road wheels on torsion arms to give a smoother ride. First of two pilot models, designated T24, was delivered in October 1943 and proved so successful that the Ordnance Department immediately authorised a production order for 1,000 vehicles which was later raised to 5,000. Cadillac and Massey-Harris undertook production, commencing March 1944 and these two plants between them produced 4,415 vehicles (including SP variants) by the war's end. In each case production supplanted M5 series vehicles.

The 75mm M6 gun was adapted from the heavy aircraft cannon used in the Mitchell bomber, and had a concentric recoil system which saved valuable turret space. The T24 was standardised as the Light Tank M24 in May 1944. First deliveries of M24s were made to American tank battalions in late 1944, supplanting M5s, and the M24 came into increasing use in the closing months of the war, remaining as standard American light tank for many years afterwards.

Parallel to the need for a new light tank was the desire to produce a standard chassis as the basis of the so-called "Light Combat Team"—a complete series of tanks, SP guns, and special purpose tanks all based on one chassis so greatly simplifying maintenance and production. The many variants produced to meet this concept are given below. Each had identical engine, power train, and suspension to the M24.

VARIANTS

M24: Production light tank (see text).

M19 Gun Motor Carriage: Produced for the AA Command, this vehicle was originally designated T65E1 and built as a development of the T65 GMC (qv) with a twin 40mm M2 AA mount set at the hull rear and the engines moved forward to the hull centre. Design (by the Ordnance Department) commenced in mid 1943 and 904 vehicles were ordered in August 1944 when the design was standardised as the M19. By the war's end, however, only 285 had been completed. M19s were standard US Army equipment for many years post-war. Crew: 6; weight 38,500lb; height 9ft 9½in; elevation −5° to +85°; stowage 336 rounds, 40mm.

CHAFFEE

M41 Howitzer Motor Carriage: Prototype for this vehicle was the T64E1, a development of the T64 HMC (qv) which had been based on M5A1 light tank components. The T64E1, however, featured the components of the "Light Combat Team" and was similar in layout to the M19, with centrally-mounted engines and the gun, a 155mm M1 howitzer, at the rear firing forward. It had a manually operated recoil spade and a folding crew platform. Unofficial name for this vehicle was "Gorilla". Standardised as the M41 HMC, in May 1945, 250 of these vehicles were ordered but only 60 were completed by the war's end. The M41 HMC was standard US Army equipment for many years post-war. Details as for M24 except: Crew: 12 (8 carried in accompanying ammunition carrier); weight: 42,500lb; length: 19ft 2in; trench crossing: 9ft; stowage: 22 rounds; range: 96 miles; elevation: +45° to −5°; traverse: 17° left to 20° right; speed: 30mph.

M37 Howitzer Motor Carriage: Intended to supplement or replace the M7 HMC (qv) a new design based on the M24 chassis was produced, resembling the M7 in general layout. Designated T76 it was standardised in November 1944 as the M37 HMC with 105mm M4 howitzer. It had the same hull arrangement (ie, rear engine) as the M24 and compared with the M7 it had greatly increased ammunition stowage and improved armour protection. American Car & Foundry were given the production contract for 448 vehicles, but only 316 were completed, most of them after the war had ended when Cadillac took over the contract. Details as for M24 except: Crew: 7; weight: 40,000lb; length: 18ft 2in; traverse: $22\frac{1}{2}$° right and left; elevation: +45° to −10°; stowage: 90 rounds.

T38 Mortar Motor Carriage: This was a project to use the M37 HMC in the mortar carrying role. The 105mm howitzer was removed and the embrasure plated over. A 4·2in mortar was carried and fired from the fighting compartment. The project was cancelled in August 1945 when it became apparent that the war would end before the vehicle could go into service. A pilot model was completed.

T77E1 Multiple Gun Motor Carriage: This was a proposed AA tank development initiated in 1943 to mount a specially designed quad ·50 cal machine gun turret on the M24 chassis. The turret was developed by the USAAF and featured remote control for the guns. Pilot vehicle, designated T77, was completed and tested at APG in July 1945. As a result of trials a computing sight system was added to the turret and the vehicle was re-designated T77E1. With the cessation of hostilities in September 1945, the project was abandoned.

M24 with swimming device: This was tested in the fall of 1944 and consisted of pontoons attached fore-and-aft to give flotation with grousers added to the tracks to give propulsion in the water, the idea being to allow the standard M24 to "swim" ashore from landing craft. Once ashore, the pontoons were jettisoned. This device was not used operationally. Designation for the device was M20.

DEVELOPMENT VEHICLES AND OTHER PROJECTS:
T24E1 was the T24 pilot model fitted with a Continental R-975 engine and (later) a longer 75mm gun with muzzle brake. As the M24 was satisfactory with Cadillac engines no further development took place with this prototype. T24 was the prototype for the M24 series; identical except for lack of vision cupola and minor details. T6E1 was the projected recovery vehicle of the "Light Combat Team", cancelled at war's end. T81 was a project to mount

237. T24 was prototype for M24 series. Note absence of vision cupola.

238. T24E1 was same vehicle with Continental engine and longer 75mm gun.

239. Standard production M19 Gun Motor Carriage.

240. T64E1 was prototype for M41 Howitzer Motor Carriage which was identical in appearance. Rear view shows recoil spade raised.

241. Standard production M37 Howitzer Motor Carriage.

242. T38 Mortar Motor Carriage, sole pilot model.

a single 40mm AA gun and twin ·50 cal MG on the T65E1 (M19) chassis. T78 was a projected improved version of the T77E1. T96 was a projected mortar motor carriage with 155mm T36 mortar. T76 was prototype (1943) of M37 HMC. There were also numerous post-war projects beyond the scope of this book.

243. T77E1 Multiple Gun Motor Carriage was a sophisticated AA tank uhich did not pass prototype stage.

British Service

A small number of M24s were supplied to Britain in 1945 and remained in service for a short time after the war. In British service the M24 was called Chaffee and this name was subsequently adopted in the U.S. Army.

Specification

Designation: Light Tank M24
Crew: 5 (commander, gunner, loader, driver, co-driver-radio operator)
Battle weight: 40,500lb
Dimensions: Length 18ft (16ft 4½in excluding gun) Track width 16in
Height 8ft 1½in. Track centres/tread 8ft
Width 9ft 8in
Armament: Main: 1 × 75mm gun M6
Secondary: 2 × ·30 cal Browning MG (one co-axial), 1 × ·50 cal AA MG
Armour thickness: Maximum 25mm
Minimum 9mm
Traverse: 360°. Elevation limits: +15° to −10°
Engine: Cadillac Twin 44T24 petrol 110hp each
Maximum speed: 35mph
Maximum cross-country speed: 25mph (approx)
Suspension type: Torsion bar
Road radius: 100 miles
Fording depth: 3ft 4in
Vertical obstacle: 3ft
Trench crossing: 8ft
Ammunition stowage: 48 rounds, 75mm
3,750 rounds, ·30 cal MG
Special features/remarks: Could be fitted with dozer blade as required. All-welded construction, homogenous armour. Highly efficient and successful design combining the virtues of speed, reliability, simplicity, ruggedness, and hitting power for a tank of its class and size.

244. Standard M24 shown in British service, where it was called the Chaffee, named for General Adna R. Chaffee, first chief of the US Armored Force.

245. M24 light tank fitted with M20 swimming device for trials.

246.

4·5in Gun Motor Carriages, T16 and T16E1: The T16 design was proposed in May 1941 and it was envisaged mounting the gun on a chassis made from components of the M3 and T7 light tanks. Cadillac carried out detailed design, and two pilot models were produced in late 1942, but based on lengthened M5A1 chassis with an extra bogie inserted on each side and the engines moved to the centre. First pilot model mounted the 4·5in gun, but the second had a 155mm howitzer and was designated T64(qv). The T16 was tested at APG and by the Artillery Test Board, the latter asking that the T24(M24) chassis be utilised for the mount in place of the M5A1. The Ordnance Department approved of this and gave the designation T16E1 to the T24-based design. The Armoured Board rejected the employment of the 4·5in gun, however, and both the T16 and T16E1 were cancelled in January 1944. The T16 is shown. T16E1 was not completed.

247.

40mm Gun Motor Carriage, T65: Designed for AA Command, this was another Cadillac design on the lengthened M5A1 chassis (see T16, plate 246). After trials, 1,000 production vehicles were requested by AA Command in February 1943, but as with the T64 (plate 248) the vehicle was redesignated on the "Light Combat Team" (T24) chassis as the T65E1. It was put into production as the M19, described and illustrated with the M24 series.

248.

155mm Howitzer Motor Carriage, T64: This was the second T16 pilot model (plate 246) armed with the 155mm howitzer and redesignated. Trials in 1943 showed this to be a good reliable design, but the Armoured Force commended that the vehicle be re-designed on the basis of the T24 (M24) chassis, since M5A1 production was scheduled to cease in favour of the M24. The Ordnance Board therefore developed a second design, T64E1, on the "Light Combat Team" (T24) chassis in August 1943 and the T64 was dropped. The T64E1 was put into production as the M41 HMC, described and illustrated with the M24 series.

249.

20mm Multiple Gun Motor Carriages, T85 and T85E1: These were alternative AA tank designs with different quad 20mm mounts respectively on the same Cadillac M5A1-based chassis as the T65, with later adaptation to the T24 chassis also contemplated. Completed in June 1944, neither went beyond trials status. (T85E1 shown).

250. Standard production M2 medium tank.

THE National Defense Act 1920 had placed tanks and tank development under the control of the Infantry and it was as an infantry support weapon that medium tanks were mainly developed in the twenties and thirties. Only a few designs were produced in this period, the T4 medium tank of 1935-36 being the only one to actually achieve limited production status. In 1938 Rock Island Arsenal designed a new vehicle which dispensed with Christie's convertible wheel and narrow hull track idea featured in the previous medium tanks. Designated the T5 the new tank was radically different, based closely on the layout of the Light Tank M2 (qv) and designed to use as many of the components as possible for reasons of economy and standardisation. The same Continental radial air-cooled engine was used, the transmission was similar, and the suspension units—vertical volute spring type—were the same. Its infantry support function was emphasised by its armament of six ·30 cal Browning machine guns, sited to give all round fire, and a central turret with 37mm gun. A full width "barbette" supported the turret and a flexible machine gun was mounted in each corner. Two fixed forward firing machine guns were mounted in the sloping hull front. The engine was at the rear and drive in the front. The weight was 15 tons.

Trials of the T5 showed that it was under powered with the Continental 250HP engine, and it was re-engined with a Wright 9 cylinder radial engine of 350HP which proved satisfactory. As rebuilt, the vehicle was known as the T5 Phase III (Phase II was an alternative suggestion for a different engine installation which was never carried out). Trials of the T5 Phase III were completed in June 1939 and the design was standardised as the Medium Tank M2, production of 15 vehicles commencing at Rock Island Arsenal in August 1939. Production vehicles had two additional machine guns fitted on the turret sides, with alternative mounts on the hull top.

An improved version of the M2 was developed in 1940 which was similar in all respects to the original design except that it had a wider vertical sided turret, armour thickness increased from a maximum of 25mm to 32mm, wider tracks, a supercharger fitted to the engine to increase its rating to 400HP, an armour cover added to the mantlet, flaps for the vision ports, improved sighting arrangements, detail changes in the suspension, and splash rails added to the hull front. Standardised as the Medium Tank M2A1, it was proposed to put this vehicle into large-scale production when the National Munitions Program was adopted on June 30, 1940, as a result of the sharp turn of events in Germany's favour in the war in Europe. Mass production of the Light Tank M2A4 (qv) was already underway by American Car & Foundry and for medium tank production it was planned to build a special plant, Detroit Tank Arsenal, which Chrysler had undertaken to erect and operate on behalf of the US Government. On August 15, 1940, a contract for 1,000 M2A1 medium tanks was placed with Chrysler, production at the rate of 100 a month to start within a year at the as yet unbuilt Arsenal.

However, the PzKw III and PzKw IV, the latter with a 75mm gun, which the Germans had been using in the invasion of France and the Low Countries, had already rendered the M2A1 medium tank, with its 37mm gun technically obsolete. On June 5, 1940, the Chief of Infantry suggested to the Ordnance Department that US medium tanks should also mount 75mm guns. Just after the M2A1 had been ordered, General Chaffee, Chief of the Armored Force (newly-formed on July 10, 1940) held a meeting with Ordnance Department representatives at Aberdeen Proving Ground, Ma, to finalise future requirements. At this important August meeting the main point agreed was the urgent need for a 75mm gun, but this could not be fitted in the small turret of the M2A1 and time would be required for development work on mounting a weapon of this calibre in a new turret. It was therefore resolved to compromise in view of the urgency of the matter and design an interim vehicle, based closely on the hull, layout, and mechanical specification of the M2A1, but with a 75mm gun in a limited traverse mounting in the right side of the hull sponson. Meanwhile work would proceed on the problem of mounting the 75mm gun in a fully traversing turret.

Basis for the new design with hull-mounted 75mm gun was the Medium Tank T5E2 which was the old T5 Phase III prototype rebuilt between March and May 1939, with a 75mm pack howitzer as an experimental self-propelled carriage. On August 28, 1940, the M2A1 contract with Chrysler was accordingly cancelled and the new vehicle, which was to become the Medium Tank M3 (qv) was substituted, even though it was yet to be designed.

As an interim measure, while the Detroit Arsenal was being built, an order for 126 M2A1 medium tanks was placed with Rock Island Arsenal and 94 of these were actually built between November 1940 and August 1941. By June 1941, however, the first M3 medium production lines were starting up, and the M2A1s were used only for training and trials purposes at various armour centres in the United States.

VARIANTS

M2: First production type, 1939 (see text).

M2A1: Second production type, 1940-41 (see text).

M2 with E2 Flame-gun: Test vehicle, 1941, for flame-throwing trials. Long flame-gun mounted in place of 37mm gun and fuel containers carried on rear hull.

DEVELOPMENT VEHICLES: T5 (later T5 Phase I) was prototype for series, 1938. T5 Phase II was projected alternative engine installation in same vehicle—not carried out. T5 Phase III was same vehicle re-engined with Wright radial engine, July 1938, becoming production prototype for M2. T5E1 was a pilot model fitted with Guiberson diesel engine and later used for trials with experimental fittings and twin 37mm guns. T5E2 was T5 Phase III vehicle converted with modified hull front and fitted with 75mm howitzer in right sponson; this layout was subsequently adopted for the M3 medium tank. In November 1940, one of the M2 mediums was used to test the installation for the prototype of the British M3 medium tank turret at Rock Island Arsenal.

251. Medium Tank T5E2 showing 75mm howitzer in right sponson and range finder in small turret on hull top. Note modified hull front.

SPECIFICATION

Designation: Medium Tank M2 and M2A1
Crew: 6 (commander, driver, gunners (4))
Battle weight: 38,00lb (M2), 47,040lb (M2A1)
Dimensions: Length 17ft 6in; Height 9ft 3in (M2: 9ft 4½in); Width 8ft 6in; Track width 14in (M2: 13in); Track centres/tread 6ft 9in
Armament: Main: 1 × 37mm M6 gun
Secondary: 8 × ·30 cal Browning MG
Armout thickness: Maximum 32mm (M2: 25mm)
Minimum 9·5mm
Traverse: 360° (main turret). Elevation limits: —
Engine: Wright radial, 9 cylinder 400hp (M2: 350hp)
Maximum speed: 26mph
Maximum cross-country speed: 17·2mph
Suspension type: Vertical volute
Road radius: 130 miles
Fording depth: —
Vertical obstacle: 2ft
Trench crossing: 7ft 6in
Ammunition stowage: 200 rounds, 37mm
12,250 rounds, ·30 cal MG
Special features/remarks: Hull was part riveted, part welded; turret was welded. Angled plates on rear fenders were intended to deflect rounds from rear machine guns down into enemy trenches. Easiest distinguishing features between M2 and M2A1 are splash plates on hull front and armoured mantlet in latter vehicle. Outclassed and obsolete at time of production, the M2A1 was nonetheless an important development vehicle since its chassis formed the basis for the M3 and M4 medium tanks.

252. Medium Tank T5 Phase III, prototype for the production M2.

253. Medium Tank T5E1. Note twin 37mm guns.

255. Medium Tank M2 used to test M3 turret installation, November, 1940.

254. Medium Tank M2 with E2 flame-gun for flame-throwing trials.

256. Rear detail view of a Medium Tank M2.

257. Medium Tank M2A1.

258. Medium Tank M3 pilot model, typical of the standard production M3.

EVOLUTION of the Medium Tank M3 from the M2 medium series has been recounted in the M2 section. The speed with which the M3 was designed, developed, and put into production was probably unmatched in the history of armoured fighting vehicles. Crucial to its production in vast numbers was the building of Detroit Tank Arsenal at Center Line, Michigan, which was expressly planned for the building of medium tanks. In September 1939 when war was declared in Europe, the Ordnance Department already had plans to contract large scale tank production to heavy engineering firms and, in fact, the first American tank to be built in quantity, the M2A4 light (qv) was turned out by one of these firms, American Car & Foundry.

Events in Europe in May/June 1940, which gave rise to the adoption of the new American National Munitions Program, showed that tanks—especially medium tanks —would be needed in far greater numbers than had been forseen the previous October when the light tank building programme was initiated. In fact, nearly 2,000 medium tanks were needed in the next 18 months according to US Army estimates which made existing orders for less than 400 light tanks seem puny by comparison. William S. Knudsen, President of GMC, was the member of the National Defense Advisory Commission responsible for co-ordinating industry to American defence needs. He advised that heavy engineering firms, used to comparatively slow and small output of such items as locomotives and cranes, would not have the capacity or expertise to turn out tanks on the vastly increased scale which, in June 1940, was now required. In Knudsen's view tank production was analogous to automobile production, and, except for manufacture of armour plate, there was no reason why the automotive industry, rather than the heavy engineering industry, should not become the prime producers of tanks, making them from the ground up all under one roof. While the Ordnance Department did not entirely agree with this view, they did concede that further tank production facilities were necessary and that the automotive men had the expertise for large volume output.

Since tank production raised problems not encountered in automobile making, Knudsen suggested that a purpose-built plant should be established in Detroit and arranged that Chrysler would build and operate this plant on behalf of the American Government. Thus was the beginning of Detroit Tank Arsenal. On August 15, 1940, the contract was signed with Chrysler for an initial order for 1,000 M2A1 medium tanks (qv) which was cancelled thirteen days later in favour of the M3 medium. Events moved fast from then on. Building of the plant commenced on a 100 acre site outside Detroit in September 1940. The building was 1,380ft long and 500ft wide.

Concurrently, Rock Island Arsenal were working on the M3 medium design with Chrysler engineers in attendance to devise plant and production equipment as design proceeded. Final M3 design work was completed in March 1941 by which time construction of the huge Arsenal building was almost finished, the whole operation taking just six months. Meanwhile the Ordnance Department contracted with two major heavy engineering firms, American Lcocomotive and Baldwin Locomotive, for 685 and 535 M3 mediums respectively. At all stages of design Rock Island Arsenal consulted with engineers and designers from the contractors concerned and there were also informal discussions with the members of the British Tank Commission, which had arrived in USA in June 1940 to place contracts for American-built tanks for the British Army. The latter were able to suggest detail improvements in the light of combat experience in the European war.

The three contracting firms all produced pilot models of the M3 in April 1941 and by August full-scale production had started in all three plants, American Locomotive, Baldwin, and Detroit Arsenal. Production of the M3 medium, and its variants, continued until December 1942, by which time a grand total of 6,258 M3 series vehicles had been turned out. In August 1941, also, Pressed Steel and Pullman each received contracts for 500 M3s from the British Commission. In October 1941, when the M4 medium design (qv) was standardised, the M3 was reclassified "substitute standard" and in April 1943 when M4s were in full service, the M3 was declassified to "limited standard", finally being declared obsolete in April 1944.

The Medium Tank M3 was dimensionally similar to the M2A1 medium and had the same Wright radial air-cooled gasoline engine and vertical volute suspension. The 75mm M2 gun (M3 in later models) was in a limited traverse mount in the right sponson and a 37mm gun was carried in a fully traversing turret off-set to the left. Maximum armour thickness was 56mm. Turret and sponson were cast and the rest of the hull was riveted though changes were made in subsequent variants as detailed separately. As originally designed the M3 had side doors and a commander's cupola, though again there were subsequent changes. Most important innovation of all, however, was the installation of gyro-stabilisers for both the 75mm and 37mm guns allowing the vehicle to fire with accuracy while on the move. This same equipment was also fitted in the M3 series light tanks (qv) from this time, (mid 1941) on. Power and hand traverse were provided for the turret, and periscope sights were fitted for both guns. Total weight of the M3 medium was 30 (short) tons.

Production variants and special purpose developments were numerous and these are detailed below with their distinguishing features:

SPECIFICATION

Designation: Medium Tank M3, M3A1, M3A2, M3A3, M3A4, M3A5, Grant I, Lee I, etc.
Crew: 6 (commander, driver, loaders (2), gunners (2))
Battle weight: 60,000lb (except where noted)
Dimensions: Length 18ft 6in (M3A4: 19ft 8in) Track width $16\frac{1}{2}$in Track centres/tread 6ft 11in
Height 10ft 3in (Grant I: 9ft 4in)
Width 8ft 11in
Armament: Main: 1 × 75mm gun M2 or M3
1 × 37mm gun M5 or M6
Secondary: 3–4 × ·30 cal Browning MG
Armour thickness: Maximum 37mm
Minimum 12mm
Traverse: Turret–360° Sponson–15° each side
Elevation limits: 75mm: +20° to –0° 37mm: +60° to –7°
Engine: See variants list
Maximum speed: 26mph (except where noted)
Maximum cross-country speed: 16mph (approx)
Suspension type: Vertical volute
Road radius: 120 miles (M3A3, M3A5, 160 miles)
Fording depth: 3ft 4in (M3A3: 3ft)
Vertical obstacle: 2ft
Trench crossing: 6ft 3in
Ammunition stowage: 46 rounds, 75mm
178 rounds, 37mm
9,200 rounds, ·30 cal MG
Special features/remarks: Interim type while M4 medium was developed but widely used by the United States, Britain and other allied nations. Sponson-mounted 75mm gun was major limitation to its fighting potential but consequent roomy hull proved useful for later special purpose conversions. Grant I was used only by British.

PRODUCTION VARIANTS

M3: Initial production type from April/August 1941 onwards. Riveted hull, side doors, and Wright radial engine,

259. Grant I which was an M3 with turret for British requirements. Note rear overhang.

260. M3A1 medium tank, distinguished by its cast hull. M2 75mm gun.

261. One of the twelve M3A2 medium tanks built.

262. M3A3 medium tank with welded up side door and counter-weight on M2 75mm gun to balance gyro-stabiliser.

Continental R-975 of 340HP. Built by Detroit Arsenal (3,243), American Loco (385), Baldwin (295), Pressed Steel (501), and Pullman (500). Total output, ceasing in August 1942, was 4,924.

M3A1: This was identical mechanically to the M3 but had a cast, instead of riveted, hull. Built only by American Loco who had the casting facilities, 300 were produced from February-August 1942. Late production vehicles had side doors eliminated and escape hatch in hull floor.

M3A2: This was mechanically identical to the M3 but had an all-welded hull. This innovation had been authorised by the Ordnance Department in September 1941, mainly to reduce weight but also because it was superior to riveting in armoured vehicles. Baldwin commenced producing the welded M3A2 in January 1942, but after only 12 were built a new engine installation was introduced (see below) from March 1942.

M3A3: All-welded hull as M3A2 but fitted with twin General Motors 6-71 diesel engines of 375HP. Weight increased to 63,000lb. Top speed increased to 29mph. Baldwin built this variant from March-December 1942, producing a total of 322. Side doors welded up or eliminated on later vehicles.

Some M3 and M3A1 models were fitted with Guiberson diesel motors due to shortage of Wright Continental motors, and these were designated with the suffix "(Diesel)".

M3A4: This model was identical to the M3 but was fitted with the Chrysler A-57 Multibank 370HP engine. This was made up of five automobile engines coupled together on a common drive shaft and was devised by Chrysler, again to alleviate the shortage of Continental engines. Built only by Detroit Arsenal from June-August 1942, 109 vehicles were produced. Weight was increased to 64,000lb and the hull was lengthened by 14in to 19ft 8in overall to accommodate the longer engine. A longer chassis and track were also necessary. Side doors eliminated.

M3A5: As M3A3 but with a riveted instead of welded hull. Built by Baldwin from January-November 1942, 591 were produced. Side doors welded up or eliminated up on late production vehicles.

All late production vehicles had the longer M3 75mm gun irrespective of model.

SPECIAL PURPOSE VARIANTS

Mine Exploder T1 (for M3 Medium Tank): This was a device consisting of a twin disc roller unit and a single disc roller unit pushed and trailed respectively for attachment to the M3 medium. Developed in early 1942, it was originally envisaged for the M2A1 medium. It was unsatisfactory in service.

M3 with E3 Flame-gun: Flame-throwing device developed from E2 as fitted on M2 medium (qv). Trials only, 1942. Flame-gun replaced 37mm gun in turret and 75mm gun was removed.

M3 with E5R2-M3 Flame-gun: Portable flame projector device supplied as a kit for quick fitting in the field in place of cupola machine gun. For details see Light Tank M3A1.

Shop Tractor T10: US-built version of British designed CDL (qv, below). Pilot model on M3. 355 production vehicles, most converted from M3A1s, May-December 1943, by American Loco. Not used in combat.

Heavy Tractor T16: This was a standard M3 tank with the turret and sponson removed and a winch fitted at the rear,

263. M3A4 medium tank with side doors eliminated and M3 75mm gun—an example of a very late production M3 series vehicle.

264. M3A4 with counter weight on M2 75mm gun and welded up side doors.

265. M3A5, view showing hull layout. Note welded up side door.

266. The M3A5 in British service was designated Grant II; this vehicle is shown in Burma, 1945.

intended to haul artillery. Prototype only, early 1942, it proved unsuitable owing to lack of stowage space for gun crew and equipment.

Tank Recovery Vehicle T2 (M31): Trials of the T16 tractor suggested the use of the M3 chassis in the recovery role. Designated T2, this was a standard M3 with guns removed and replaced by dummy barrels, plus the addition of a rear mounted boom and winch, and tool boxes. Produced as a "limited procurement" conversion in September 1942, the T2 was redesignated M31 in September 1943 and classified "limited standard". M31B1 was same conversion on M3A3 and M31B2 was based on the M3A5. Winch had 60,000lb pull.

Full-Track Prime Mover M33: This was the M31 ARV further converted as a tractor for the 155mm gun, 1943-44. Boom and turret were removed and air compressor and outlet lines were added to operate brakes on gun carriage. A ·5 cal AA MG mount was added on hull top. Pilot conversion was designated T1. **Tractor M44** was similar but with a cupola on the hull sponson.

3in Gun Motor Carriage T24: Developed in September 1941, this was a project to mount a 3in gun on the M3 medium chassis to produce a tank destroyer. Turret and sponson were removed from a standard M3 as was the hull top. Trials showed the vehicle to be too high and the mounting was too complicated to mass-produce. Project cancelled in March 1942.

3in Gun Motor Carriage T40 (M9): This was an improvement on the T24 GMC, to mount a redundant M1918 3in AA gun in a low angle mount on the M3 chassis for the tank destroyer role. In December 1941 it was proposed to build 50 such vehicles to use up 50 guns thought to be available. After trials the 50 vehicles were ordered in April 1942 and the vehicle was designated M9 GMC in the "limited standard" category. However, only 28 guns were found to be available and it was thought that the M10 GMC would be in production before the M9 GMC order could be completed. Project was cancelled in August 1942.

40mm Gun Motor Carriage T36: Project to mount a 40mm AA gun and its direction equipment on the M3 medium chassis, proposed by the AA Board in October/November 1941. Test showed this vehicle to be too complex and undergunned for its size and the project was cancelled.

(Note: gun and howitzer motor carriages on the M3 medium chassis which actually reached production status are described separately at the end of the M4 medium tank sections.)

DEVELOPMENT AND EXPERIMENTAL VEHICLES: M3E1 was vehicle with Ford GAA engine to test installation for M4A3 medium tank. M3A1E1 was test vehicle with triple 6 cylinder Lycoming engines. M3A5E1 was test vehicle to try Twin Hydra-matic transmission. M3A5E2 was test vehicle with normal Hydra-matic transmission. An M3A4 was tested in 1942 with experimental trailing idler hinged to rear bogie. Another was tested with horizontal volute spring suspension.

BRITISH SERVICE

A British Tank Commission had been sent to the United States in June 1940 when there was a grave shortage of tanks for the British Army and much of the British tank strength had been lost in France at the time of the Dunkirk evacuation. The Commission were charged with the task of procuring American types for British service and arranging for the production of British tank designs in the United States. With the defeat of Britain then seemingly very possible, the

267. Standard M3 in British service where it was designated Lee I. Sand shields were added by British to M3 series vehicles in Middle East.

268. Standard production Tank Recovery Vehicle T2(M31); British designation Grant ARV I.

269. Grant ARV was a British conversion for the recovery role. Some had a dummy turret but were otherwise similar.

270. Grant CDL, late vehicle with dummy barrel on turret.

National Defense Advisory Commission opposed any production of British designs, since all productive capacity was needed for the American tank programme. In the medium tank field, therefore, the British Commission was left only with the choice of ordering the M3. In October 1940 they placed direct contracts with Baldwin, Lima, and Pullman for M3 mediums for Britain. The M3s built and paid for by Britain (ie, the initial order) had a new cast turret to meet British requirements that called for the radio equipment to be mounted in the turret rear instead of in the hull as in the original design. This turret was longer than the original M3 turret with prominent rear overhang and a pistol port in each side. The cupola was eliminated to reduce the silhouette and the turret itself was lower, reducing the vehicle's overall height by about a foot. This version was called the Grant by the British (after General Ulysses S. Grant) and the 200 vehicles ordered were all shipped to the 8th Army in the Western Desert starting early in 1942. For the big Gazala battle on May 27, 1942, 167 Grants formed the bulk of the equipment of 4th Armoured Brigade and at last gave the British a tank with superior fire power to any opposing German AFV. For the first time the British had a tank in service which could out-range the German panzers and had the added feature of a "dual purpose" capability with a 75mm gun which could fire AP or HE as necessary, the latter in the close support and indirect fire roles. The M3 Grant gave a welcome boost to the morale of British tankmen, helped to inflict a big reverse in the fortunes of the Afrika Korps, swung the balance of tank power both qualitatively and quantitatively in favour of the British from then on, and, as an after effect, started the British off on the quest to design a similar 75mm dual purpose gun for mounting in British-built tanks. It was, indeed, at that time, the most important new addition to the British armoury.

On March 11, 1941, the Lend-Lease Act was ratified which made munitions of all kinds available to Britain and to others of the Americans' allies. Standard M3 mediums were thus also supplied to the British Army who gave these vehicles the name Lee (after Robert E. Lee). By June 1942 a further 250 M3 medium tanks had arrived in Egypt for the 8th Army and by the time of the Battle of Alamein in October 1942, a total of nearly 600 M3 series medium tanks had been delivered under a combination of "cash and carry" and Lend-Lease terms. In June 1942, also a maintenance unit had been set up in a depot near Cairo where US Army personnel familiarised British crews with the M3 (and later M4) mediums.

A small number of M3 series mediums were shipped to Britain for training and special purpose use and conversions, but the main bulk of this type in British service were used in the Middle East. When the M4 replaced the M3 in this theatre, the M3s were shipped to Burma where they equipped British units previously using Matildas, Stuarts, Valentines, and other obsolete types. Some also went to the Australians at this period.

Variants and special purpose types in British service are detailed individually below:

Grant I: M3 with turret to meet British requirements. Distinguished from other variants by lack of cupola (see main text). Original British "cash and carry" contract.

Grant II: British designation for M3A5 medium tank. Original US type turret.

Lee I: British designation for M3 medium tank.

Lee II: Designation for M3A1 medium tank.

Lee III: Designation for M3A2 medium tank, though none were delivered to the British.

271. Early Grant CDL, using the M3A1 chassis and lacking dummy gun.

272. T1 Mine Exploder on M3 medium tank. Vehicle has turret traversed while operating to protect 37mm gun.

273. M3 with E3 Flame-gun. Note 75mm gun removed.

274. Grant Scorpion III was a British conversion. Note 75mm gun removed.

275. T24 3in Gun Motor Carriage.

Lee IV: Designation for M3A3 fitted with Continental engine.

Lee V: Designation for diesel-engined M3A3.

Lee VI: Designation for M3A4 medium.

Grant ARV: British conversion, 1943, of obsolete Grant I or II, with guns removed and towing winch and limited recovery equipment installed. Two versions: (1) Few with dummy turret, (2) With turret removed and replaced with hatch and twin Bren AA MG mount.

Grant ARV I: British designation for standard US T2(M31) TRV in British service, 1944-45.

Grant Command: Grant fitted with extra radio equipment for use of senior officers. Few only, with 37mm gun sometimes removed or replaced by a dummy barrel.

Grant Scorpion III: For full description of Scorpion equipment and development, see Matilda and Valentine sections. The Scorpion III equipment was produced at the start of the Tunisian campaign, January 1943, by REME workshops in response to a request for a lighter version of the original tyoe to enable the carrying tank to move faster. Only a few were made, for fitting to the Grant, and the 75mm gun was removed to give clearance for the rotor frame. There was a counter weight at the rear.

Grant Scorpion IV: This was the Scorpion III improved by the addition of a second Bedford engine at the left rear to provide a more powerful drive for the rotors.

Grant CDL: For development of the CDL tank, see Matilda CDL. A complete CDL unit (1st Tank Brigade) with Matildas was formed in the Middle East early in 1942 but not committed to action. In early 1943 when M4s had replaced Grants in 8th Army, it was decided to convert some of the redundant vehicles to the CDL role as replacements for the out-dated Matildas. Grants so altered retained the 75mm gun, so keeping an offensive role, and had the turret replaced with the armoured searchlight housing which included a ball-mounted machine gun in the front face. Later vehicles had a dummy 37mm barrel added to the turret front. The Brigade was earmarked for operations in NW Europe, 1944, but CDL tanks were in the event only used in small numbers to illuminate the night crossings of the Rhine and Elbe in early 1945. Other CDL tanks were sent to the Far East in 1945, but never used. Meanwhile, US Amored Force officers had witnessed CDL demonstrations and 355 vehicles were converted to this role in late 1943 to equip six American tank battalions for operations in Europe. The American version was designated T10 Shop Tractor, the name being chosen to disguise the true role. (See American M3 Medium section).

276. M33 Prime Mover towing 155mm gun, Italy 1944, M44 was similar but with covered cupola over right sponson.

277. T40 3in Gun Motor Carriage, standardised as M9 but never put into service.

278. T36 40mm Gun Motor Carriage.

279. M3A5E1 medium tank was an experimental vehicle with Twin Hydra-matic transmission. Note extended rear hull to accommodate this.

280. Early production Medium Tank M4 (British: Sherman I). Note vision blocks in glacis plate, three-piece bolted nose casting, and narrow M34 gun mount.

DESIGN of the Medium Tank M3 had been undertaken as a development of the M2A1 on the clear understanding that it was to be considered as an interim design to get a tank with 75mm gun armament into production and service as soon as possible. While design was carried out on the M3 the Armored Force Board drew up requirements for its successor with a 75mm gun in a fully traversing turret. Final M3 working drawings were completed in March 1941, and Rock Island Arsenal offered the Armored Force Board five suggested schemes for the M4 at a meeting the following month. The most straightforward scheme was selected, which entailed using the M3 medium chassis, suspension, power unit, transmission, and other mechanical parts, unchanged, and providing a completely new hull top, either cast or welded, with a central turret mounting the 75mm gun. The 37mm gun was discarded but a machine gun cupola was to be retained on the new turret. Doors were provided in each side of the hull as in the M3. Designated Medium Tank T6, this vehicle was built in wooden mock-up form in May 1941 for Armored Force Board approval, and a pilot model, with cast hull and detail changes which included elimination of the cupola, was completed at Aberdeen Proving Ground on September 19, 1941. It is tempting to suggest that the T6 was influenced, if not copied to some extent, from the Canadian Ram (qv) in view of its similarity. Documentary evidence (and the chronology of events) disprove this, however. An early production Ram was sent from Montreal Locomotive works in July 1941 for tests at Aberdeen Proving Ground which lasted until October that year, but APG's report on the Ram was concerned only with its comparison to the M3 and offered no comment on its relevance to the T6 design.

Meanwhile the German invasion of Russia in July 1941 indicated that American involvement in the war in Europe would increase in the year ahead. On President Roosevelt's personal orders tank production schedules for 1942, provisionally set at 1,000 medium tanks a month, were doubled. To achieve this, additional production facilities were required and Pacific Car & Foundry, Fisher, Ford, and Federal Machine & Welder were added to the list of plants earmarked to build the new medium tank. In October 1941, the T6 was standardised as the Medium Tank M4 and plans were made to introduce the M4 on to the production line, in those plants building M3s, at some convenient point early in 1942. This would mean that M4 medium tanks would be built at a total of 11 plants in 1942. A major proposal was that a second purpose-built tank production plant be built on the lines of Detroit Arsenal. In September 1941 Fisher were asked to erect and operate such a plant at Grand Blanc, Michigan. Building of Grand Blanc Tank Arsenal, designed from the start to turn out M4s, was started in January 1942 and tank production commenced the following July, though Fisher had, meantime, commenced building M4s in one of their existing plants.

The M4 pilot model was built by Lima Locomotive works in February 1941, differing from the T6 principally in the elimination of the hull side doors. Full production in three plants, Lima, Pressed Steel, and Pacific Car & Foundry started the following month, all these initial production types being cast hull vehicles, designated M4A1. By the autumn of 1942 all other plants in the programme were in full production, and in October 1942, at the Battle of Alamein, the first M4 mediums went into action with British forces. The M4 series was the most widely produced, most widely used, and most important of all tanks in service with American, British, and allied forces in World War II. While not necessarily the best Allied tank in qualitative terms, and certainly inferior in armour and hitting power to the best German and Soviet tanks, the M4 medium tank (popularly known by its British name of Sherman) had the virtues of simplicity of maintenance, reliability, speed, ruggedness, and an uncomplicated design. These were most important factors for a vehicle being mass-produced in commercial plants with no background of military experi-

281. Medium Tank T6 which was the prototype vehicle for the M4 series. Note M2 gun with counterweight, M3 suspension, and side doors.

282. Early production M4A1 with vision slots in hull front, M3 type suspension, M34 gun mount, M3 gun. This particular vehicle was the finest production vehicle in the M4 series.

ence in peace-time, for use largely by conscript troops with training time limited by the needs of war. In terms of cost-effectiveness, the M4 Sherman was supremely suited to the needs of the hour, a fact reflected in the total output of more than 40,000 tanks (and associated AFVs) based on the M4 chassis in the years 1942-46. Shermans were used by every allied nation in every armour role on every fighting front. A full outline of production types and variants is given below.

The Medium Tank M4 had the same basic chassis as the M3 medium, with vertical volute suspension, rear engine, and front drive. Apart from very early models, the bogies were altered, however, so that the return rollers were set behind, instead of on top of, the spring units. Hull was either welded, cast, or welded with cast/rolled nose, as detailed in descriptions of the different models, while the 75mm gun was set in a simple cast turret and provided with a gyro-stabiliser as in the M3. Initially the engine was a Continental R-975 air-cooled radial type, but an ever-persistent shortage of this Wright-built power unit (which was essentially an aircraft engine and needed as such by the aero industry) forced the adoption of alternative engines, giving rise to the main production variants. The M4 Sherman had a five man crew, could fire AP or HE shot, had a maximum armour thickness of 50mm (more on reworked and late models), had controlled differential steering, weighed from 33 (short) tons, gross, and had a top speed of 24-30mph according to model. There were numerous detail changes in models during their production life and these are detailed for individual types.

283. Early production M4A2, May 1942, showing twin fixed MG in hull front and all other early M4 series features.

284. Late production M4 with one-piece nose M34A1 gun mount, and appliqué armour fitted.

SPECIFICATION

Designation: Medium Tank M4 series
Crew: 5 (commander, gunner, loader, driver, co-driver-hull gunner)
Battle weight: 66,500lb (M4A1), 69,000lb (M4A2), 68,500lb (M4A3)
Dimensions Length 19ft 4in (average) Track width 16½in
Height 9ft Track centres/tread 6ft 11in
Width 8ft 7in
Armament: Main: 1 × 75mm M3 gun
Secondary: 2 × ·30 cal MG
1 × ·50 cal MG (AA)
Armour thickness: Maximum 75mm (turret, 50mm (hull))
Minimum: 12mm
Traverse: 360°. Elevation limits: +25° to −10°
Engine: See variant details
Maximum speed: 24–29mph
Maximum cross-country speed: 15–20mph (approx)
Suspension type: Vertical volute
Road radius: 100–150 miles (according to engine)
Fording depth: 3ft (3½ft in M4A4/A6)
Vertical obstacle: 2ft
Trench crossing: 7ft 5in (8ft M4A4/A6)
Ammunition stowage: 97 rounds 75mm
4,750 rounds ·30 cal
Special features/remarks: M4A1 distinguished from rest by cast hull. M4A4 and M4A6 had longer hull and longer tracks (see variant details). Later vehicles had sand shields and wide M34A1 gun mounts and, often, were reworked with appliqué armour. Vehicles with 105mm howitzer, details generally as above for appropriate model but carried 66 rounds of 105mm ammunition and were fitted to tow ammunition trailer.

SHERMAN

Production Models

M4: Continental R-975 engine and welded hull. Early vehicles had a three-piece bolted nose, and vision slots (later eliminated) for driver and hull gunner. They had the narrow M34 gun mount. Later production vehicles had a one-piece cast nose and wide M34A1 gun mount. Very late production vehicles (late 1943) had a combination cast/rolled hull front. The M4 was built by Pressed Steel (1,000, July 1942-August 1943), Baldwin (1,233, January 1943-January 1944), American Loco (2,150, February-December 1943), Pullman (689, May-September 1943), Detroit Arsenal (1,676, August 1943-January 1944). Total output: 6,748.

M4A1: As M4 but with cast hull. This was the first model to go into production. Very early production models had M3 type bogie units, M2 75mm gun with counter weights, and twin fixed machine guns in hull front. These guns, plus hull vision slots, were soon eliminated and the M3 75mm gun was introduced after the first few vehicles. Nose was changed from three-piece bolted to one-piece cast, and M34A1 mount and sand shields were added in later vehicles. The M4A1 was built by Lima (1,655, February 1942-September 1943), Pressed Steel (3,700, March 1942-December 1943), Pacific Car & Foundry (926, April 1942-November 1943). Total output: 6,281.

M4A2: Second type in production, this vehicle differed from the M4 in having twin General Motors 6-71 diesel engines to overcome the shortage of Continental gasoline engines. Production changes effected corresponded with those described for the M4 with the exception of the cast/rolled hull front which never appeared on this model. Very early vehicles had spoked road wheels (as did very early M4A1s) but these were soon abandoned in all models for solid disc type with embossed spokes to simplify production. The M4A2 was built by Fisher/Grand Blanc (4,614, April 1942-May 1944), Pullman (2,737, April 1942-September 1943), American Loco (150, September 1942-April 1943), Baldwin (12, October-November 1942), Federal Welder (540, December 1942-December 1943). Total output: 8,053. Used only by USMC. Most went into Lend-Lease stocks.

M4A3: Fifth type in production, this model had a welded hull and was fitted with a 500HP Ford GAA V-8 gasoline engine specially developed for the vehicle. This was the production type most favoured by the US Army and most of these were retained for American use, Lend-Lease deliveries being mainly concentrated on other models. Production changes as for M4 except that this model had the one-piece cast nose for its whole production life. Ford built 1,690 M4A3s from June 1942-September 1943 before ceasing tank production for other munitions work. M4A3 production was then taken over by Grand Blanc from February 1944-March 1945, and numerous improved features were incorporated, including a vision cupola for the commander, a loader's hatch, 47° hull front, and "wet stowage" for the ammunition. This was the most advanced M4 series vehicle with 75mm gun, the improvements being in line with those in up-gunned variants described below. Total output of improved M4A3: 3,071.

M4A4: Fourth type in production, this was built only by Detroit Arsenal following on M3 production from July 1942-September 1943. All M4A4s had the three-piece bolted nose, while early vehicles had the M34 gun mount and vision slots in the hull, later eliminated. Late vehicles had the M34A1 gun mount which offered better protection to the mantlet. The M4A4 had the Chrysler WC Multibank engine, made from five commercial automobile engines on a common drive shaft. This was another expedient to

285. Standard production M4A1 with one-piece nose, steel tracks, vision slots eliminated from hull, and M34 gun mount.

286. Late production M4A2 with one-piece nose, M34A1 gun mount, sand shields and appliqué armour.

overcome the engine shortage. To hold this engine it was necessary to lengthen the rear hull to give an overall length of 19ft 10½in. This 370HP unit gave the vehicle a top speed of 25mph. Total M4A4 output: 7,499.

M4A5: This was the American designation allocated but not used for the Canadian-built Ram (qv) which was also developed from the M3 medium series. One Ram I was tested by the US Army (see main text).

M4A6: Final basic model to enter production this was essentially the M4A4 hull and chassis with a newly-developed Caterpillar RD-1820 radial diesel engine of 450HP. It has a lengthened hull like the M4A4 to accommodate the engine and correspondingly lengthened track. Major improvement in the hull was a cast/rolled front to give increased armour production. Detroit Arsenal produced the M4A6, following straight on after M4A4 production ceased. Between October 1943-February 1944, 75 M4A6 vehicles were produced after which orders for this model were cancelled due to production difficulties with the engine and the need to rationalise engine types in use. The same cast/rolled hull front was also used on the last of the Detroit-built M4s, however.

Design Improvements

User experience led to numerous design improvements being suggested by the Armored Force and incorporated into M4 series vehicles by the Ordnance Department. Foremost among these were the need for a more powerful gun and better protection. Improved features introduced to the M4 series were as follows.

(1) **76mm gun:** To increase firepower the Ordnance Department developed the 76mm gun M1 and M1A1, starting in July 1942. Tests showed that the existing M4 series turret was too small to accommodate the extra length of this weapon and the turret of the T20/T23 medium tank (qv) was adopted and suitably modified. The 76mm gun installation was standardised and introduced in production lines from February 1944 and vehicles so fitted were available in time for the Normandy landings and subsequent combat in NW Europe. Suffix "(76mm)" indicated vehicles with this gun. A modified 76mm gun M1A1C or M1A2 with muzzle-brake was later introduced.

(2) **Protection:** Fire hazard from hits in the engine, ammunition bins, and fuel tanks was the major shortcoming in the M4 series due to the relatively thin armour. Expedient measures to combat this were the addition of appliqué armour plates on hull sides adjacent to ammunition bins and fuel tanks, plus further appliqué armour welded on hull, and sometimes turret, front. Field modifications by crews included the use of sandbags on hull front and the welding of spare track shoes in vulnerable spots. In some instances large armour shields or concrete were added to hull fronts. Major design change to overcome the problem was the introduction of "wet stowage" (glycerine-protected) ammunition racks in 76mm-armed and late 75mm-armed vehicles. Howitzer-armed vehicles had internal armour plates on ammunition racks.

(3) **Suspension:** Introduction of the heavier gun and other improvements increased the vehicle's all-up weight with adverse effects on the ride. In 1943 a new wider (23in) T80 track was developed with centre line guides. At the same time a new horizontal volute spring suspension (HVSS) was designed to replace the vertical volute type. Bogies were dimensionally similar, but there were four wheels to each

SPECIFICATION

Designation: Medium Tank M4 series with 76mm gun
Crew: 5 (as M4 with 75mm gun)
Battle weight: 70,000–72,800lb (approx) (varied with basic model)
Dimensions: Length 24ft 3in (over gun)
20ft 4in (over sand shields)
Height 9ft 9in
Width 8ft 9½in (with HVSS approx 15in wider)
Track width 16½in (23in with HVSS)
Track centres/tread 6ft 11in
Armament: Main: 1 × 76mm gun M1, M1A1, M1A1C, or M1A2
Secondary: As earlier models
Armour thickness: Maximum 62mm
Minimum 12mm
Traverse: 360°. Elevation limits: As vehicles with 75mm gun
Engine: See variant details
Maximum speed: 29–24mph
Maximum cross-country speed: 15–20mph (approx)
Suspension type: Vertical volute or HVSS
Road radius: 85–100 miles (according to engine)
Fording depth: 3ft
Vertical obstacle: 2ft
Trench crossing: 7ft 6in
Ammunition stowage: 71 rounds 76mm
6,250 rounds ·30 cal MG
Special features/remarks: Late models M4A1, M4A2, or M4A3 only, all with "wet stowage" for ammunition racks, 47° hull front, and T23 type turret with vision cupola. M1 gun had plain muzzle, M1A1 had small counterweight, and M1A1C and M1A2 had muzzle-brake. Could be seen with vertical volute suspension or HVSS. Full details in main text. British Firefly had generally similar characteristics (not HVSS or "wet stowage") but hull machine gun was eliminated, some had British type vision cupola, and modified original turret was used with rear extension. 42 rounds 17pdr were carried.

287. Standard production M4A3 (Ford-Built).

288. Early production M4A4 with vision slots in hull and M34 gun mount. Later models had vision slots eliminated and M34A1 gun mount.

289. Standard production M4A6; note cast/rolled hull front.

290. Standard production M4A1(76mm) with "wet stowage" and M1A1 76mm gun.

and the horizontal springs were tougher than the old vertical springs. Return rollers were now mounted on the hull sides. HVSS was designed so that any wheel could be replaced without removing the complete bogie concerned. HVSS was introduced from mid 1944. Vehicles so fitted had track covers on the hull sides due to increased width of the tracks.

The original track for the M4 series had incorporated rubber blocks, but to overcome rubber shortage steel tracks were designed, in two different patterns. They were interchangeable with the rubber tracks and all three types could be seen on M4s. To improve ride in muddy conditions, grousers could be fitted to the outer edge of these earlier tracks, giving increased effective width. With T80 tracks grouseres were not supplied, though later developed.

(4) **105mm Howitzer:** For the close support role a 105mm howitzer was contemplated and two M4A4s were modified to take this in November 1942. This was standardised in 1943 for production.

(5) **Miscellaneous:** Other improvements included better electrical wiring, and other internal detail changes, a new 47° hull front to simplify production (it also improved frontal protection), larger access hatches for driver and co-driver, a loader's hatch, and the provision of a vision cupola for the commander replacing the rotating hatch ring originally fitted.

Late Production Types

Arising from the improvements noted, the following M4 series models appeared:

M4 (105mm): As M4 with new armament. 800 produced at Detroit Arsenal, February-September 1943.

M4 (105mm) HVSS: As M4 (105mm) but with modifications 2, 3, 4, 5 above. 841 built by Detroit Arsenal, September 1944-March 1945.

M4A1 (76mm): As M4A1 but with modifications 1, 2, 5 and (later) 3, above. 3,396 built by Pressed Steel, January 1944-June 1945.

M4A2 (76mm): As M4A2 but with modifications 1, 2, 5 and (later) 3, above. Built by Grand Blanc (1,594, June-December 1944) and Pressed Steel (21, May-June 1945). Total output: 1,615.

M4A3 (76mm): As M4A3 but with modifications 1, 2, 5 above. Built by Detroit Arsenal (1,400, February-July 1944) and Grand Blanc (525, September-December 1944). Total output: 1,925.

M4A3 (76mm) HVSS: As M4A3 but with modifications 1, 2, 3, 5 above. 1,445 built by Detroit Arsenal, August-December 1944.

M4A3 (105mm): As M4A3 but with modifications 2, 4, 5 above. 500 built by Detroit Arsenal, April-August 1944.

M4A3 (105mm) HVSS: As M4A3 but with modifications 2, 3, 4, 5 above. 2,539 built by Detroit Arsenal August 1944-May 1945.

M4A3E2 Assault Tank: This was a compromise design proposed in early 1944 for use in ETO as a heavily armoured tank for the support of infantry when it was realised that the T26E1 heavy tank could not possibly be ready for service before early 1945. Since the US Army had no other heavy tank in service, it was decided to modify the M4A3 for the heavy role. Additional armour was added to all hull surfaces, giving a maximum thickness of 100mm. A new heavy turret was designed with frontal armour of 150mm but the 75mm gun was retained. Total weight as modified was 42 (short) tons and grousers were permanently fitted to the track in an

291. Later production M4A1(76mm) HVSS with M1A1C gun.

292. Late production M4(105mm) HVSS. Note commander's vision cupola. M4A3(105mm) HVSS was of similar appearance.

293. Standard M4A2(76mm) with M1A1 gun. Note 47° hull front modifications clearly visible.

294. Later M4A1(76mm) HVSS with M1A1C gun. Compare with plate 293.

295. M4A3E8 was prototype for the M4A3(76mm) HVSS, production vehicles being identical to this prototype. Compare with plate 287 showing original M4A3 appearance.

attempt to improve the ride. Top speed 22 mph. Procurement of 254 vehicles, classified "limited standard", was authorised and these were built by Grand Blanc in May-June 1944 and rushed to Europe. In service, a few vehicles were refitted with 76mm M1 guns taken from damaged M4 (76mm) tanks. The M4A3E2 was known unofficially as "Jumbo".

SPECIAL PURPOSE VARIANTS

Tank Recovery Vehicle M32: This was a modification of the standard M4 for the armoured recovery role. Turret and gun were replaced with a fixed turret and a 81mm mortar to fire smoke. 60,000lb winch was fitted in fighting compartment and a pivoting A-frame jib was mounted on the hull. Extra tow-points, tow bars, blocks, and other recovery items were also added. Weight: 62,000lb; speed: 24mph; length: 19ft $1\frac{1}{4}$in; A-frame: 18ft. Other details as M4.

Tank Recovery Vehicle M32B1: As above but based on M4A1 (cast hull) chassis.

Tank Recovery Vehicle M32B2: As above but based on M4A2 chassis.

Tank Recovery Vehicle M32B3: As above but based on M4A3 chassis. Later vehicles had HVSS.

Tank Recovery Vehicle M32B4: As above but based on chassis of M4A4.

M32 series was standardised in September 1943. Pilot model, built by Lima, was designated TRV T5.

Full-Track Prime Mover M34: This was the M32B1 modified by the removal of the A-frame, winch, and other recovery gear to provide a tractor for heavy artillery guns. Air compressor and air pipes were added for braking tow. In service 1944.

M4 Dozer: A few M4s were fitted with dozer blades and hydraulic hoists taken from Caterpillar D-8 dozers for work on the Italian front soon after the landings in 1943. These

296. Standard production M4A3(76mm) with M1A1 gun. This preceded the M4A3(76mm) HVSS. Compare with plate 295.

297. M4A3E2 showing new turret and added hull armour.

298. Full-track Prime Mover M34.

299. M4 Dozer, extemporised type with blade and hydraulic gear from D–8 Dozer.

300. M4A1 fitted with M1 dozer blade. Note armoured cover over lift gear.

301. M4A1 Dozer with Engineer Corps, showing turret removed.

302. Tank Recovery Vehicle M32.

303. Tank Recovery Vehicle M32B3, showing deep wading trunking in position.

304. M4 Mobile Assault Bridge.

proved so successful that a special dozer blade M1 was designed and produced as a standard fitting for the M4 series in 1944. A second, wider, version, the M1A1, was also made, to fit vehicles with HVSS. The dozer blade was attached to the centre bogie each side and worked hydraulically from the M4's power traverse. A few M4s were permanently fitted as dozers for the Engineer Corps and these had the turret removed.

M4 Mobile Assault Bridge: Field modification in Italy, this consisted of a double track bridge supported for travelling by an A-frame jib. The bridge was driven into the gap to be bridged and dropped into place, the carrier vehicle, an old M4, could not recover the bridge unaided. Only a small number were built. Vehicle was heavily ballasted at the rear.

M4 with Cullin Hedgerow Device: Vehicle fitted in the field, Normandy, June 1944, with "prong" cutters made from old beach defences to facilitate passing through hedges and undergrowth of "bocage" country. Seen on any variant, including SP vehicles based on M4 series chassis. Also used by British.

Mine Clearing Devices on M4 Series Vehicles

Roller, flail, and plough equipment was evolved for mine clearing in US Army service. Rollers and flails detonated mines by surface pressure while ploughs excavated mines for defusing. Many types were tried, but only a few of these were produced for service, usually on a "limited procurement" basis. Mine Exploders T1 and T2 were developed for use with the M3 medium and M3 light tanks respectively and are described in the relevant sections devoted to these vehicles. The following were developments for use with the M4 medium tank series:

Mine Exploder T1E1 (Earthworm): A development of the T1 (qv), this equipment differed in that the trailing disc roller unit was brought to the front so that the forward unit now formed a tricycle layout. The discs were now made of armour plate. It was designed for use with the Tank Recovery Vehicle M32, the vehicle's boom being used to support the disc roller unit and lift it as required. The T1E1 saw limited use. It was developed and produced in 1943. Weight, 18 (short) tons.

Mine Exploder T1E2: This was a modified version of the T1E1 with the disc roller unit being reduced once more to two forward units, each with seven discs instead of the six discs per roller used in the T1 and T1E1. Experimental only, 1943.

Mine Exploder T1E3 (M1) (Aunt Jemima): Also developed in 1943, this device consisted of two roller units each of five 10ft diameter steel discs driven by chain and gearing from the front sprocket of a suitable modified M4 series tank. 75 were built. Placed in service in Normandy and Italy as a "limited standard" item of equipment (M1) in 1944, it was the most widely used of American mine exploders. M1A1 was improved version with solid discs lacking cut-outs. Second tank was sometimes needed to push Jemima equipment.

Mine Exploder T1E4: A 1944 development, this consisted of 16 discs in a single heavy-frame unit, pushed by a suitably modified M4 series tank and suspended by a heavy reinforced curved boom.

Mine Exploder T1E5: Developed in July 1944 from the T1E3 (see above), this equipment differed in having smaller diameter discs, six to each roller unit, with a central support frame instead of an outside frame. Gear driven from the propelling tank's sprockets, each roller unit had independent movement. Experimental only.

Mine Exploder T1E6: As T1E3 (see above) but with serrated edges to discs. Experimental only.

Mine Exploder T2E1: Similar in appearance to Light Mine Exploder T2 used on the M3 light tank (qv), this was a scaled up type developed for the US Marine Corps for use with the Tank Recovery Vehicle M32, utilising the vehicle's boom to support and drag the weights. It proved impractical and work on the project was abandoned in October 1943.

305. Mine Exploder T1E1 (Earthworm) with M32 propelling vehicle.

306. Mine Exploder T1E3(M1) (Aunt Jemima) with M4A1 propelling vehicle.

307. Mine Exploder T1E4 with M4A3(76mm) propelling vehicle.

Mine Exploder (T2) Flail: American designation for British Crab I equipment (qv, British M4 section) tested in USA. A small number were used by the US Army in NW Europe, 1944-45.

Mine Exploder T3: Designed at the end of 1942, following British use of the Matilda Scorpion (qv), this device utilised the British rotor and chains, but took drive from the vehicle's engine and had American-designed arms. It proved unsatisfactory, and development was stopped late in 1943.

Mine Exploder T3E1: This was the T3 re-built with longer arms and a sand-filled rotor to improve performance. It still proved unsatisfactory and was cancelled.

Mine Exploder T3E2: A further development of the T3E1 in which the rotor was replaced by a steel drum of larger diameter. Work on this was terminated at the war's end in 1945.

Mine Exploder T4: US designation for British Crab II (qv) tested and used in small numbers by the US Army.

Mine Excavator T4: Simple plough device evolved in late 1942 which featured an angled blade with curved cross-section and prongs on the bottom of the plough. This was attached rigidly in front of an M4 tank. It proved impractical under test and was abandoned.

Mine Excavator T5: This was similar to the T4, but improved by having the plough changed to a V-shape instead of being straight. A modified version of this was designated T5E1.

Mine Excavator T2E2: Developed in late 1943 from the T5E1, this incorporated the arms and hydraulic lift gear from the M1 dozer so that the plough could be raised or lowered.

Mine Excavator T6: Another design based on an angled V-shape plough, this and two modified versions, T6E1 and T6E2, proved unsatisfactory due to inability to control blade depth. As a result it was suggested by the Armored Force Board that the best features of this design be incorporated in the T5E2 based on the T5E1 (above).

Mine Excavator T5E3: This was the T5E2 with features from the T6. It was ordered on a "limited procurement" basis in June 1944. In this vehicle the angled plough was mounted on the front of the complete M1 dozer blade assembly.

Mine Exploder T7: Developed in late 1943, this consisted of a frame carrying small rollers, each made up of twin discs and pushed ahead of an M4 tank. It proved unsatisfactory on test and the project was abandoned.

Mine Exploder T8: This device consisted of steel plungers carried on a pivoted frame in front of an M4 tank, arranged with gearing to beat up and down on the ground as the vehicle moved forward. Tests at APG showed the idea to be impractical and steering of the vehicle adversely affected. It was abandoned in March 1944. Nicknamed "Johnnie Walker".

Mine Exploder T9: This was a heavy spudded 6ft roller attached to a frame and pushed ahead of an M4 tank. Tested at APG it proved extremely difficult to manoeuvre and work on the T9 was cancelled in favour of a lighter design, essentially similar, the T9E1. Due to its lighter weight, however, this device tended to "jack-knife" under certain conditions, and also occasionally failed to explode mines in its path. The whole project was cancelled in September 1944.

Mine Exploder T10: The National Defense Research Council suggested, early in 1944, a remote controlled mine exploder device consisting of three roller units, articulated in tricycle layout and controlled by a following tank. This was developed into a self-propelled unit, consisting of the roller units as first envisaged, but powered and driven from a modified M4 hull/turret mounted above the articulated tricycle unit. Designated T10, it was tested at APG, but proved to be unwieldy and the project was cancelled in the fall of 1944.

Mine Exploder T11: This was a M4A4 with 6 spigot mortars set to fire forward. Experimental only.

308. Mine Exploder T3 utilised British Scorpion rotor and US arms.

309. Mine Exploder T3E2 was a development of the T3.

310. Mine Excavator T5E2.

311. Mine Exploder T8.

Mine Exploder T12: This was a spigot mortar launcher platform carrying 23 mortars, mounted in the fighting compartment of a turretless M4. The spigots were set at various angles to spread the mortars over an area directly ahead of the vehicle, thus exploding mines laid in its path. Tested at APG, the T12 proved most effective, but work on the project was cancelled in December 1944 since it appeared that a similar installation using rockets instead of mortar bombs would be simpler.

Mine Exploder T14: This was an M4 tank with added belly armour and a heavy duty track and suspension designed to explode mines simply by passing over them and relying on its heavy armour for immunity. Experimental only.

Mine Resistant Vehicle T15, T15E1, T15E2: Resulting from experience with the T14, three M4 tanks were modified by the removal of the turrets, addition of extra belly and side armour, and the fitting of very heavy duty tracks and suspensior units, reinforced by armoured brackets. The three differed in detail but were of similar appearance and construction. Work on these started in September 1944, but the project was abandoned when the war ended.

Snake Equipment for M4: This was a 1943 experiment using the British method of pushing a snake explosive charge ahead of the tank and detonating the charge to clear a path through mines. It was not adopted.

Improvised devices: Several equipments were developed by units in the field, usually based on the flail principle. Typical of these was a US Marines device which utilised the attachments, hydraulic lift, and other parts from the M1 dozer kit and carried chain flails and a drive mechanism on a welded-up framework. Another device, built by 89th Ordnance Bn in Tunis, March 1943, featured a flail on a fixed girder framework mounted ahead of the tank with a drive shaft from the tank's engine.

FLAME-THROWER DEVICES ON M4 SERIES VEHICLES

E4R2-5R1, E4R3-5R1, (M3-4-3) Flame-guns: These were simple devices, varying in detail, which were mounted in place of the hull machine gun, being supplied as a kit for fitting as necessary in the field.

E4R4-4R 5-6RC Flame-gun: This was a periscope type flame-projector, similar to the above, which also fitted in place of the hull machine gun. Fuel container was carried in the vehicle. It gave a slightly longer range than the M3-4-3 type.

POA Flame-thrower: This was a Pacific theatre improvisation using a US Navy Mk I flame-thrower with the projector tube fitted inside an old 105mm howitzer barrel with the breech removed. Fuel container carried inside turret.

POA-CWS 75-H1: As above, but utilising the barrel of an old 75mm M3 gun.

POA-CWS 75-H2: As above but with projector tube attached to the right side of the 75mm barrel so that the vehicle retained its gun armament.

Many POA-fitted vehicles used by USMC in the Pacific had detachable wooden covers on the hull sides to prevent attachment of magnetic charges by Japanese suicide troops.

E6-R1 Flame-gun: Another portable flame-thrower device which came in kit form for fitting in the periscope aperture in the assistant driver's hatch.

E7-7 Flame-gun: Short projector which replaced main gun, with fuel carried in hull. Same equipment could be fitted to M3A1 and M5A1 light tanks.

312. Mine Exploder T9E1; T9 was of similar appearance but heavier.

313. Mine Exploder T10.

314. Mine Destroyer and Demolition Vehicle. M4A4 fitted with Spigot Mortar attachment.

315. Mine Resistant Vehicle T15E1.

316. Improvised flail device used in Pacific by USMC, utilised M1 dozer fittings.

317. M3–4–3 Flame-gun mounted in M4A1(76mm).

318. POA–CWS 75 H1 Flame-thrower on USMC M4A3 fitted with wooden side covers.

319. M4 Crocodile of US 2nd Armored Division.

Ronson Flame-gun: This was the Canadian Ronson flame-thrower device fitted to a M4. Few vehicles so fitted by USMC. For further details see M3 light tank section (Satan).

M4 Crocodile: This was British Crocodile equipment (qv, Churchill section) fitted to the M4 but with the fuel pipe taken over the hull instead of under the belly. Four M4s were so converted for US use in NW Europe in late 1944. They were used by the US 2nd Armored Division.

E1 Anti-personnel Tank Projector: Devised in 1945, this consisted of four small projectors fitted to the hull of a M4A3 and fired, either individually or simultaneously from inside the hull to ward off suicide troops attacking the vehicle at close quarters. Intended for use in the Pacific theatre, it was still undergoing trials at the war's end and was subsequently dropped. It was also known as the "Scorpion".

Other flame-thrower equipments in M4 series vehicles were the E13R1-13R2 and E20-20 both installed in M4A3s, but not completed until late 1945.

ROCKET LAUNCHER MOUNTS ON M4 SERIES VEHICLES

While numerous rocket launcher mounts were developed for fitting to M4 series vehicles, very few saw operational use or reached production status.

Rocket Launcher T34 (Calliope): This consisted of 60 4·6in rocket tubes mounted in a frame above the turret. The two bottom sets of 12 tubes each could be jettisoned if necessary on all variants except the M4A1. The mount was traversed with the tank turret and elevated by a rod linked to the gun barrel. The Calliope was a "limited procurement" weapon, developed in 1943 and first used by 2nd Armored Division in France in August 1944. This weapon saw limited combat use until the end of the war.

Rocket Launcher T34E1: As T34 but with 14 tubes in two bottom projector units.

Rocket Launcher T34E2: Similar in appearance to the T34, but longer, the T34E2 held 60 7·2in rockets and the entire mount could be jettisoned if necessary in an emergency. This mount saw limited combat use, 1945.

Rocket Launcher T39: A mount of enclosed box construction with doors over the tubes. It held 20 7·2in rockets. Experimental only.

Rocket Launcher T40(M17) (Whiz-bang): This rocket launcher held 20 7·2in rockets in a box-like frame and was elevated hydraulically from the 75mm gun controls. The entire mount could be jettisoned if required, and the rockets could be fired singly or in salvoes. This "limited procurement" weapon was classified "limited standard" and saw some combat use in 1944-45.

320. Rocket Launcher T34 (Calliope) on M4. US 80th Divn, March 1945. Note typical camouflage and addition of grousers on track.

321. El Anti-personnel Tank Projector on M4A3(76mm).

322. Rocket Launcher T40(M17) (Whiz-bang) showing method of loading.

325. Rocket Launcher T105 mounted on M4A1 test vehicle.

323. Rocker Launcher T40 (short version); note side door.

326. Rocket Launcher T72. Note appliqué armour on reworked M4A2 carrier vehicle.

324. Rocket Launcher T76 mounted on M4A1 test vehicle.

327. M4 under test with M19 flotation device which was similar to that tested on the M24 (qv).

Rocket Launcher T40 (short version): Experimental version of the above with shorter rocket tubes and 75mm gun removed and replaced by elevation mechanism for launcher. Access door for crew added in side of vehicle which was an M4A2.

Rocket Launcher T72: Similar to T34 but with very short tubes. Not used operationally.

Rocket Launcher T73: Similar to T40 but held only 10 rockets. Not used in combat. Experimental only on M4A1.

Rocker Launcher T76: This was a M4A1 with a 7½in rocket tube replacing the 75mm gun. Had an opening in turret front around the mounting to allow gases to escape on firing. Reloaded from inside turret. Experimental only, 1944. Same weapon mounted on M4A3 HVSS was designated T76E1.

Rocket Launcher T105: A single 7·2in rocket projector in box-like case mounted in M4A1 in place of 75mm gun. Developed from T76, August 1945. Did not proceed past trials stage.

Multiple Rocket Launcher T99: Two small box-like launcher mounts, each holding 22 4·5in rockets, mounted each side of turret for vehicle with 76mm gun. Few produced 1945; also fitted experimentally to M26 heavy tank.

Self-propelled Gun Mounts on M4 chassis

Demolition Tank T31: Experimental vehicle produced in 1945 on a M4A3 HVSS chassis with a 105mm howitzer in a fabricated heavily armoured turret with a 7·2in rocket projector on each side. Prototype only.

Multiple Gun Motor Carriage T52: A design initiated in July 1942 by Firestone for either two 40mm or a single 40mm flanked by two ·50 cal machine guns in a ball-type traversing mount to provide an AA tank on the M4 chassis. The latter alternative arrangement was chosen and built for trials. Its traverse proved too slow for low flying aircraft and the project was terminated in October 1944.

90mm Gun Motor Carriage T53 and T53E1: Development of a 90mm AA gun on a M4 chassis was started in July 1942. The gun was to be mounted on the rear of the chassis and the engine would be mounted centrally. A pilot model on an M4A4 chassis was duly constructed at Detroit Arsenal by Chrysler and this was tested at APG. As a result of these tests 500 vehicles were authorised, subject to changes which included moving the gun to the centre of the vehicle and the engine back to its original position. As re-designed, the vehicle was designated T53E1 and was to serve as a dual purpose AA/anti-tank vehicle. Tests by Tank Destroyer Command showed it to be no better than the M10 and AA Command also rejected the vehicle, so the production order, and the entire project, was cancelled in May 1944. The T53E1 had outriggers on the bogies for emplacement in the AA role.

(Note: gun and howitzer motor carriages on the M4 medium chassis which achieved production and service status are described separately at the end of the M4 medium tank section.)

Development and Miscellaneous Experimental Vehicles

M4E1(i): was M4A1 fitted with 76mm gun T1 in standard turret to investigate possibility of up-gunning, July/August 1942. Turret later modified with rear bustle to overcome

328. Demolition Tank T31.

329. Multiple Gun Motor Carriage T52.

330. Gun Motor Carriage T53E1 in final form; note retracting jacks on bogies.

331. M4A1 test vehicle with T1 76mm gun in original turret was sometimes unofficially designated M4E1.

332. M4E6 test vehicle with T23 type turret and M1A1 gun, plus cast/rolled hull front.

333. M4E5 was test vehicle for 105mm howitzer installation.

space problem caused by 76mm breech. Later a T23 turret was substituted. M4E1 was unofficial (and incorrect) designation.

M4E1(ii): was M4A4 fitted with Caterpillar RD-1820 diesel engine to test installation for M4A6 production vehicle.

M4E2: was M4A4 with 24in wide T80 track and early HVSS suspension as development vehicle (1943) for improved suspension. Another vehicle was fitted with HVSS developed from the original vertical volute type.

M4E3: was M4A4 fitted with a Chrysler V-12 engine for test.

M4E4: was M4 fitted with 24in wide tracks and experimental torsion bar suspension.

M4E5: was M4 fitted with 105mm howitzer as prototype for vehicles with this weapon, early 1943.

M4E6: 1943, was test vehicle for T23 turret with M1A1 gun and "wet stowage" for ammunition. It also featured the cast/rolled hull front.

M4E7: was M4A1 fitted with Ford GAA engine in place of the Continental R-975 engine to investigate feasibility of Ford engines for this model. Not adopted.

M4E8: was an early M4A3 fitted as a test vehicle for HVSS and 76mm gun, the latter being fitted in the original turret.

M4E9: was M4 with spaced out suspension and extended end connectors (grousers) in an attempt to improve vehicle's ride.

334. M4A2E4 was a test vehicle for torsion bar suspension and 23in tracks.

335. M4A3E1 was typical test vehicle, in this case for a Spicer torque converter. Note added weights for trial running.

M4A1E1: was M4A1 fitted with aluminium foil insulation and air conditioning to test improved habitability for desert warfare conditions.

M4A1E2: was M4A1 fitted with a recording odograph and infra-red lights to test night-fighting techniques.

M4A1E3: was M4A1 fitted with a test installation of the Spicer Model 95 torque converter.

M4A1E5: was M4A1 with improved Continental engine, better cooling, and increased fuel capacity, all later standardised in late-production M4 series tanks with Continental engines.

M4A1E8: was prototype, early 1944, for the production M4A1 (76mm) with "wet stowage" and HVSS.

M4A1E9: was M4A1 with the same modifications as M4E9 (see above).

M4A2E1: was a 1943 trials vehicle with its twin GM diesels replaced by a GM V-8 diesel motor.

M4A2E4: was M4A2 with the same modifications as M4E4 (see above).

M4A2E9: was M4A2 with the same modifications as the M4E9 (see above).

M4A3E1: was M4A3 with the same test modification as the M4A1E3 (see above).

M4A3E2: was prototype for the heavy assault tank put into limited production and service (see M4 Late Production Types for description).

M4A3E3: was prototype to test 47 degree hull front.

336. M4A3E9 was a test vehicle for spaced out suspension and modified track with 9in grousers.

337. M4 towing armoured sledges, seen under test by British.

338. T10E1 Shop Tractor was experimental American-built CDL on M4A1. T10 was similar searchlight mounting on M3 medium tank.

339. M4A4 with Allis-Chalmers experimental suspension and wide tracks.

340. M4A4 test vehicle with early form of HVSS.

41. M4A1 Centipede with T16 half-track suspension units for trials.

M4A3E8: was 1944 prototype for production M4A3 (76mm) HVSS with "wet stowage".

M4A3E9: was M4A3 with modified test suspension as M4A9 (see above).

M4A4E1: was M4A4 with prototype gun mount for 105mm howitzer.

M4 Doozit: was M4 with M1 dozer blade with wooden platform attached for placing demolition charges against walls and other obstacles to be breached. It also carried a T40 rocket launcher. Developed by the US Engineer Corps as an assault engineer vehicle, it was not used in combat.

M4 with Assault Sledges: Experimental project to tow "train" of armoured sledges, each with an infantryman inside, for the assault role under fire. It was also tested by the British, but trials showed the idea to be impractical.

T10E1 Shop Tractor: Experimental conversion mounting CDL equipment on M4A1 chassis. Details as for T10 described in M3 medium section.

Among scores of other minor M4 experimental types were included M4A4 with Allis-Chalmers suspension (HVSS type), M4A3 with Rheem automatic gun loader, M4 with M19 swimming device, M4A1 Centipede with suspension units from the T16 half-track and numerous post-war developments (and production variants) which are beyond the scope of this book.

342. British Sherman Firefly VC–17pdr gun, modified turret, based on M4A4. Note hull MG eliminated.

British Service

ISSUED to the British, the M4 medium tank was named the "General Sherman", more often called the "Sherman", by which name it also became popularly known by other nations. The Sherman was, in fact, the most important tank in British service and more widely used than any of the British designed or British produced types from 1943-45. British built tanks with comparable 75mm "dual purpose" gun power (eg, the Cromwell IV and Churchill VII) were not available for service until the end of 1943 and not in wide service until spring 1944. The first Shermans, almost all the cast-hull M4A1 variant, were shipped to the 8th Army in the Middle East in October 1942 and about 270 of the first batch of 300 were in service at the start of the Battle of Alamein on October 24, 1942, where they supplemented the M3 mediums (Grants) to make up almost half of the British tank strength of 1,100 vehicles committed to battle. To make up the numbers of M4s available, some of the vehicles had been withdrawn from American armoured units in USA to replace vehicles lost by the sinking of one of the freighters convoying the initial batch to Massawa.

At the end of 1942, also, shipments of M4s started to Britain and there was a steady flow of these vehicles to the British, both in Britain and the Middle East from then until the war's end, all vehicles being supplied on the Lend-Lease basis. While numbers of all production variants were supplied to Britain, major deliveries were of the M4A4 type (more than 1,600 supplied to 8th Army in Italy in 1943), M4A2, M4, M4A1, and M4A3 in roughly that order of quantity. Few M4A3s were sent, since this was selected as the principal service type for the US Army. Very late production vehicles with HVSS, 76mm guns, and "wet stowage" were not delivered until late 1944, then only in very small numbers, since most of these improved types went to the US Army. Some of these late vehicles were used by the British however.

The major development in British service was the fitting of the 17pdr gun to a proportion of Shermans to provide the most powerfully armed British tank of the war. Known as the Sherman Firefly, the fitting of the 17pdr gun had been suggested in January 1943 as a safeguard against the failure of the Challenger programme (qv), this latter tank proving, on trials, to have several shortcomings. Though there was some opposition to this idea at the Ministry of Supply, the British War Office insisted on a pilot conversion being produced. This was ready in November 1943 and in February 1944 the Firefly conversion was given full priority for service following delays and uncertainties with the Challenger, which prevented it being in service in time for Operation Overlord, the Normandy landings. The Sherman Firefly was the only British tank landing at Normandy which could take on the German Tiger and Panther tanks on anything approaching equal terms and proved a most successful expedient design. Initially Sherman Fireflies were issued on the basis of one per troop, due to the shortage of 17pdr guns available for fitting in tanks. By early 1945, however, the type was in service in more generous numbers. In late 1945, a Firefly turret was sent to APG for test firing mounted on a M4A3 chassis. It was evaluated for the US Army but not adopted for service.

In British service a large number of indigenous special purpose conversions were produced on the Sherman chassis, as were many experimental types. These are all described below.

British Basic Variants

Sherman I: British designation for standard M4.

Sherman Hybrid I: British designation for late-production Detroit-built M4 with combination cast/rolled hull front.

Sherman IB: British designation for M4 (105mm)

Sherman IBY: British designation for M4 (105mm) HVSS. Delivered late 1945.

Sherman II: British designation for standard M4A1 (cast hull).

Sherman IIA: British designation for M4A1 (76mm).

343. Very early production M4A1 delivered to British (British: Sherman II). Note M2 75mm gun with counterweight, M34 gun mount, and twin fixed MG in hull.

344. M4A1 (76mm) in British service as Sherman IIA. Note stowage box added on turret rear.

Sherman IIC (Firefly): British designation for M4A1 re-armed in Britain with 17pdr gun.

Sherman III: British designation for standard M4A2.

Sherman IIIAY: British designation for M4A2 with 76mm gun, "wet stowage", and HVSS. Delivered late 1944.

Sherman IV: British designation for standard M4A3.

Sherman IVA: British designation for standard M4A3 with 76mm gun and "wet stowage".

Sherman IVB: British designation for M4A3 (105mm).

Sherman IVC (Firefly): British designation for standard M4A3 rearmed in Britain with 17pdr gun.

Sherman V: British designation for standard M4A4.

Sherman VC (Firefly): British designation for M4A4 re-armed in Britain with 17pdr gun. Most Firefly conversions were on the M4A4 chassis. Hull machine gun (and gunner) deleted in all Fireflies to increase ammunition stowage.

Sherman VII: British designation for M4A6. Few, if any, delivered to Britain.

Sherman V (Rocket): A field modification by the Coldstream Guards, Guards Armoured Division, to provide twin launchers for 60lb aircraft rockets (taken from Typhoon fighters) on sides of Sherman V turret, late 1944. Though demonstrated to senior officers, the modification was not adopted elsewhere. Some vehicles so fitted were Sherman VC.

345. British Sherman Firefly IVC–note extension on turret rear to give room for 17pdr breech. One-piece nose and appliqué armour.

346. British Sherman Hybrid Firefly IC on very late Detroit-built M4—cast/rolled hull front.

BRITISH SPECIAL PURPOSE VARIANTS

Sherman III, ARV Mk I: M4A2 with turret removed and equipped with winch in turret space, A-frame demountable jib, and other standard British ARV fittings.

Sherman V, ARV Mk I: As above but based on the M4A4 chassis.

Sherman V, ARV Mk II: British conversion of M4A4 with dummy fixed turret and dummy gun, jibs front and rear, earth spade, and 25 (long) tons winch to same standards as Churchill ARV II (qv).

Sherman II (M32B1), ARV Mk III: British designation for standard US M32B1 TRV supplied under Lend-Lease for British use.

Sherman BARV: Following a decision in October 1943 that a recovery vehicle for beach work would be required for the Normandy landings, A Sherman ARV Mk I was converted with an added welded superstructure, bilge pump, and engine intake trunking for deep wading. Tested in December 1943, it proved most successful, able to work in up to 9ft of water. 21 Army Group requested 50 (later 66) and 52 were delivered by D-Day, June 6 1944. They were towing vehicles only, since to simplify the conversion and shorten production time, winches were omitted. In postwar service the BARV was named "Sea Lion". One or two interim conversions were also built with simplified, smaller, superstructure and retaining normal wading equipment permanently installed.

Sherman Kangaroo: Employment of the RAM Kangaroo APC in NW Europe in the fall of 1944 was so successful that the 8th Army commander requested a regiment of APCs for use on the Italian front. Between October 1944 and April 1945, 75 Sherman IIIs (together with some M7s) were

347. Standard British Sherman III (US: M4A2) shown with wading trunking erected, Italy 1943.

so converted by brigade workshop units. Turret was removed, and interior gutted to give room for 10 infantrymen plus the crew of two.

Sherman Fascine Carrier: Conversion by 79th Armoured Division by removal of turret on "war-weary" vehicles, and provision of sloped framework and release gear to carry two or three fascines. Similar conversion, not used operationally, was the Sherman Crib, which had a wooden platform on the hull front, turret traversed aft, and carried a wooden crib on the platform.

Sherman Gun Tower: Conversion of old M4A2 in Italy as a towing vehicle for 17pdr anti-tank gun. Small number converted by brigade workshops by removal of turrets. Vehicle carried gun crew and ammunition.

Sherman OP/Command/Rear Link: Vehicle with extra radio equipment for use of OP officers of SP artillery regiments or senior officers of armoured formations. OPs and some others had dummy gun.

Sherman Twaby Ark: This was a bridging vehicle similar to the Churchill Ark (qv). It had trackways fitted fore-and-aft over the hull top, the turret removed, and hinged ramps at each end of the trackways which were supported by kingposts for travelling. The vehicle was driven into the ditch or river to be spanned, released the ramps and kingposts, and stayed there, considered expendable if necessary. A second vehicle was needed to assist in hoisting the ramps. Several types, differing in detail, were produced on the Sherman chassis. Sherman Arks were used mainly for trials and training, the Churchill Arks being used in combat.

Sherman Octopus: An experimental development of the Ark, this was similar but with longer ramps. Two types were produced, the first with ramps made up from girder sections, the second with longer ramps of lattice construction.

Sherman Plymouth: This was a turretless Sherman with supports on the hull top enabling a length of made-up Bailey bridge to be carried. These were used, mainly in Italy, as carrier vehicles for mobile bridges as alternatives to Churchills.

Other bridgelayer experiments based on the Sherman included trials with a SBG bridge as carried by the Churchill AVRE (qv).

Sherman DD (Duplex Drive) Tanks

Straussler DD equipment was evolved in 1941 and tested on the Tetrarch light tank (qv). It proved a most successful method of giving amphibious capability to a standard tank, and the first vehicle produced in numbers with this equipment was the Valentine DD (qv). This was an obsolescent type by 1943, however, so Valentines were used only for training crews in DD techniques and a start was made, in April 1943, to adopt the DD system to the Sherman. The vehicle was waterproofed and fitted with a collapsible canvas screen round the hull top. This was erected by rubber tubing filled by compressed air. Struts locked the screen in place. Two small propellers driven from the power take-off of the vehicle's engine provided propulsion at 4 knots through the water, and were declutched and folded away when on land. Sherman DD tanks equipped a complete brigade of 79th Armoured Division for the Normandy landings in June 1944 and were the first British tanks to land, "swimming" ashore from the LCTs of the invasion fleet. In the DD system the screen gave flotation and the tank actually hung below water level; due to the low free-board, DD tanks were easily swamped and operations with these vehicles in rough weather could be hazardous without very careful handling.

Sherman III and V DD: These were the original conversions using similar screens and fittings to the Valentine DD.

Sherman III and V DD Mk I: Improved conversion for issue to troops with strengthened top rail and inter-locking struts.

Sherman III DD Mk II: DD Mk I with detail improvements.

Sherman III and IIIAY DD Mk III: American conversions for British and (limited) American use with detail changes. Latter vehicle had HVSS and 76mm gun. In British service late 1945.

Experimental developments based on DD vehicles included: BELCH equipment evolved in 1944 to provide a water spray around DD screen to protect it from fire; Sherman V DD APC which was a standard vehicle with turret removed and replaced by mushroom-shaped armour cover for over head protection. This was intended to carry infantry ashore with the DD tanks; Sherman III DD with Rocket Egress was a DD tank fitted with aircraft JATOG No 5 Mk 1 rockets each side to assist vehicle to climb steep river banks; Sherman Topee which had pontoons fore-and-aft in addition to DD equipment to improve flotation. The Topee was a flotation device fitted and tested by the British on a Sherman DD; Sherman DD Mk II with Ginandit was a DD tank with a mechanical device for dropping mat sections ahead of the vehicle to enable it cross soft mud.

SHERMAN MINE CLEARING DEVICES

Most of the devices for mine clearing developed in the years to 1943 were adopted to fit the Sherman. In addition some new devices were evolved specifically for use with the Sherman.

Sherman Scorpion IV: This was the same equipment developed for the Grant Scorpion IV (qv) mounted on a Sherman III, North Africa, May 1943.

Sherman Pram: Experimental device of flails and rollers carried on two rotor arms. Drive came from the front sprocket by means of chains. Not developed beyond test stage.

Sherman Marquis: This had a flail assembly which could be lifted hydraulically and a rotor drive engine carried in an armoured housing in place of the turret. It was an experimental development based on the Valentine Scorpion II (qv). It was originally called Sherman Octopus.

Sherman Lobster: This was a further development from the Marquis utilising similar hydraulic rotor arms but with flail drive taken via a power take-off from the engine. The turret was retained. It was the forerunner of the Crab and developed in May, 1943.

Sherman Crab I and II: A further refinement from the Lobster, this was developed in June 1943 and put into production and a complete brigade of 79th Armoured Division was equipped with these vehicles for the Normandy landings. The Crab rotor had 43 flailing chains which beat the ground ahead of the vehicle, drive for the rotor coming via a power take-off from the engine. Beneath the rotor arms were wire-cutters for clearing barbed wire. There was a screen across the hull front and covers over the driver's and assistant driver's periscopes to offer protection against dust and earth thrown up by the flails. Crabs were organised in troops of five and were allocated for mine clearing operations as required by local commanders. Crabs were fitted with lane-marking equipment (to indicate "swept" lanes) and station-keeping lights for following vehicles. The Mk II version differed from the Mk I in having a contouring device to give better coverage of rough or irregular ground.

A Sherman BARV was also fitted experimentally with Crab equipment, but used only for trials.

Sherman with CIRD: This was identical to the CIRD equipment fitted to the Churchill (qv, for full description).

348. Sherman V of Coldstream Guards fitted with 60lb aircraft rocket and launcher.

349. Sherman ARV Mk I showing A-frame jib erected. Note twin Bren AA MG.

350. Sherman ARV Mk II with front jib erected. This was dismantled for travelling.

351. Sherman BARV shown on Normandy beaches.

SHERMAN

Sherman with AMRCR: This was identical to the AMRCR equipment fitted to the Churchill (qv, for full description).

Sherman with Centipede: This was a device of 12 small concrete rollers on parallel bars specifically intended to explode German anti-personnel "S" mines. The Centipede saw limited use. It was evolved in May 1943.

Other experimental mine clearing equipment developed for use with the Sherman included: **Porcupine,** which was a pair of spiked rollers dragged in front of the tank on a boom rigged on the nose of the vehicle; **Lulu,** which was an electrical mine detecting equipment consisting of three wooden rollers carried on booms, two in front and one behind the tank. The rollers contained mine detector coils which registered on an indicator inside the tank if the roller passed over a mine; **Jeffries, Bullshorn,** and **MDI** mine ploughs were all tried with the Sherman. For details of the Jeffries and Bullshorn types see Churchill section. MDI plough was a simpler type fitted to plates welded on the nose of a Sherman V. None of these ploughs were used operationally on Shermans. Snake and Conger equipment were also tested with the Sherman. For full description of these see Churchill section. Another explosive device tried with the Sherman was the Tapeworm; this consisted of 500 yards of 2¾in explosive-filled hose in a trailer. The trailer was towed to the edge of the minefield by a Sherman with CIRD; from there the Tapeworm was towed across the minefield by the Sherman and detonated when the tank reached the far side. The CIRD exploded any mines in the path of the tank as it advanced. The 50ft of hose nearest the tank were filled with sand to insulate the vehicle from the explosives in the rest of the hose.

SHERMAN FLAME-THROWERS

Sherman V Adder: This conversion featured flame fuel in an armoured tank at the rear of the vehicle. Fuel was led forward across the hull top to a traversing projector on the co-driver's hatch. An armoured cover protected the fuel pipe. Supply was maintained by a pump. It was originally known as the Cobra. Developed in 1944, but not used operationally.

Sherman Salamander: There were eight different versions of this all varying in installation and based on the Wasp flame-throwing equipment used in the Universal Carrier. Some had the flame-projector inside a dummy gun barrel, and others had it co-axial with, or below, the gun. Experimental only, 1943-45.

Sherman Crocodile: Four vehicles converted by British for American use. See American M4 section.

Sherman Badger: M4A2 (HVSS) with turret removed and fitted with Wasp equipment with flame-gun in place of hull machine gun. This was a Canadian development of 1945. See also Ram Badger.

Miscellaneous British experimental vehicles included: A Sherman AVRE, built for evaluation in 1943 against the Churchill AVRE which was ultimately selected for service; Sherman CDL, experimental vehicle with armoured turret as used on Grant CDL (qv); Sherman with grapnels and lines fired by rockets to clear barbed wire entanglements in an assault landing; a Sherman CIRD which was tested with Bangalore Torpedoes on the CIRD arms; a turretless Sherman carrying an assault boat to approach river crossings etc, under fire. The British also used the standard US M1 and M1A1 dozer blade. In British service, vehicles so fitted were known as Sherman Tankdozers.

There were also several other less important British experimental variants based on the Sherman chassis.

352. Sherman Kangaroo (converted Sherman III).

353. Sherman Fascine Carrier.

354. Sherman Gun Tower, hauling 17pdr anti-tank gun.

355. Typical Sherman OP/Command/Rear Link tank.

356. Sherman Octopus being crossed by Churchill VII; this is the original Octopus with girder ramps.

357. Sherman Crab II showing lane markers and station keeping lights at rear.

358. Sherman Twaby Ark showing ramps in travelling position, with kingposts rigged.

359. Sherman Octopus fitted with the later lattice ramps, shown in travelling position.

360. Sherman Plymouth showing Bailey Bridge carried on hull top.

361. Sherman DD Mk I showing canvas flotation screen partly raised and propellers rigged at rear.

362. Sherman Scorpion IV; compare Grant Scorpion, plate 274.

363. Sherman Marquis, showing flail arms raised.

364. Sherman Lobster was immediate fore-runner of Crab.

365. Sherman with CIRD shown fitted experimentally with Flying Bangalore Torpedoes which were rocket fired to clear barbed wire obstacles.

366. Sherman with AMRCR.

367. Sherman with Lulu mine detecting equipment.

368. Sherman with Jeffries plough shown folded for travelling.

369. Sherman with Bullshorn plough shown sweeping for mines.

370. Sherman with MDI mine plough. Basic vehicle is a Sherman V.

371. Sherman Topee.

372. Standard production M4(105mm) in service with a Free French unit on D-Day June 6, 1944. Note 47° hull front and 105mm howitzer.

373. Standard production M7B1 HMC showing cast one-piece nose and M4 type bogies, plus added hinged side plate.

PLANS for mounting a 105mm howitzer on the chassis of the M3 medium tank to provide self-propelled artillery support for armoured divisions were made in June 1941 as soon as the M3 was in production. Two pilot models were constructed, designated T32, based on M3 medium chassis, but with an open topped superstructure. The standard M1A2 105mm howitzer was installed with its carriage suitably modified to fit. The weapon was offset to the right of centre. The trials vehicles were successful and the design was standardised as the M7 HMC in February 1942. Changes (made in the T32) for production vehicles included modified front shields and a cupola and ring for an AA machine gun. American Loco started production in April and built 2028 in 1942. Late production vehicles had M4 type bogies with trailing return rollers; otherwise the M7 had identical chassis and mechanical specification to the M3 medium tank. A few late production M7s also had a one-piece cast nose instead of the original three-piece bolted type.

Meanwhile the M4 had superseded the M3 in production and in September 1943 it was proposed to continue M7 production on the chassis of the M4A3 medium tank. Designated M7B1, this vehicle differed from the M7 (aside from its different power plant) in having hinged side plates for added ammunition protection, a cast one-piece nose, and bogies with trailing return rollers. Pressed Steel Corporation built 826 from March 1944-February 1945, after which Federal Welder built another 127 vehicles of both M7 and M7B1 type by the war's end.

With the standardisation of the M37 HMC on the Light Combat Team (M24) chassis (qv) in January 1945, the M7 and M7B1 were reclassified to "substitute standard" and were gradually replaced by the M37 from then on. The M7 and M7B1 HMCs were standard equipment of artillery battalions in all American armoured divisions.

SPECIFICATION

Designation: 105mm Howitzer Motor Carriage M7 or M7B1
Crew: 7 (commander, driver, gun crew (5))
Battle weight: 50,634lb (M7B1: 50,000lb)
Dimensions: Length 19ft 9in (M7B1: 20ft 3¾in) Track width 16½in
Height 8ft 4in Track centres/tread 6ft 11in
Width 9ft 5¼in
Armament: Main: 1 × 105mm howitzer M1A2, M2 or M2A1
Secondary: 1 × ·50 cal MG (AA)
Armour thickness: Maximum 62mm
Minimum 12mm
Traverse: 15° left, 30° right. Elevation limits: +35° to −5°
Engine: Continental R–975 radial air-cooled (M7) Ford GAA V8 (M7B1)
Maximum speed: 25–26mph
Maximum cross-country speed: 15mph (approx)
Suspension type: Vertical volute
Road radius: 85–125 miles
Fording depth: 4ft (M7), 3ft (M7B1)
Vertical obstacle: 2ft
Trench crossing: 7ft 6in
Ammunition stowage: 69 rounds 105mm
300 rounds ·50 cal MG
Special features/remarks: Very light gun for such a large chassis, but the M7 series were an expedient type got quickly into service and enjoying the advantages of standardisation and reliability inherent in the M3/M4 medium chassis. Proved very successful in service though crew protection was limited.

British Service

In March 1942 The British Tank Mission in the United States saw the M7 pilot model and immediately requested 2,500 for British use for the end of 1942 with a further 3,000 for delivery in 1943. These demands were never met in full owing to the need to equip American forces first. Due to the serious position of the 8th Army in the Western Desert in September 1942, however, 90 M7s were sent to the British, diverted from production intended for American troops. They arrived in time to play an important part in the crucial Alamein battle and several hundred more vehicles were sent in the following months. They remained in service with 8th Army throughout the Italian campaign. In British service the M7 was designated "105mm SP, Priest". Priests also equipped some of the artillery battalions with British armoured divisions in the Normandy landings, June 1944, but these were replaced by Sextons (qv) a few days after landing, partly in order to standardise ammunition in 21 Army Group and partly to make more 105mm ammunition available to the American forces. The following British conversions for special purposes were made:

Priest Kangaroo: A total of 102 vehicles were converted to Kangaroo APCs, October 1944-April 1945, by the removal of the gun and mount, removal of ammunition bins, and plating in of the hull front. This work was carried out by various brigade workshops "in the field" and the vehicles were used (together with Sherman Kangaroos (qv)) by a specially organised APC regiment of 8th Army on the Italian front. The Priest Kangaroo carried 20 infantrymen and their equipment plus a crew of two.

Priest OP: This was a conversion of a redundant Priest, 1944, by the removal of the gun and installation of extra radio, field telephones, etc, to provide a vehicle for artillery observation officers. Appearance was similar to that of the Priest Kangaroo.

25pdr Howitzer Motor Carriage T51: This was an M7 fitted experimentally with a British 25pdr in July 1942 to meet British requirements. It was not produced, however, the British using the standard M7. The prototype conversion was carried out at APG.

374. Standard production M7 HMC, shown fitted with deep wading equipment at rear.

375. Priest Kangaroo with British troops in Italy, 1945. Note three-piece bolted nose, main distinguishing feature of the M7.

376. T32 pilot model for M7 HMC series.

377. Important American-built type delivered to 8th Army was the M7 Priest, which played a big part in the Battle of Alamein in October 1942. This vehicle is seen later, after the invasion of Sicily in July 1943.

378. Standard production M10 GMC, turret traversed aft. Note counter weights on turret rear.

FOLLOWING the successful fitting of a 105mm howitzer on the medium tank chassis, plans were made in April 1942 to mount a high velocity gun on the medium chassis to provide a complementary SP vehicle for the Tank Destroyer Command. Designated T35 this vehicle utilised an early production M4A2 tank chassis, then just available, with an open-topped low sloped turret adapted from the turret design for the T1 Heavy Tank, and the 3in gun projected for the same vehicle. However, the Tank Destroyer Board asked for a lower silhouette and angled hull superstructure, so an improved design T35E1 was drawn up, again on the M4A2 chassis, and incorporating these features. The T35E1 was modified with thinner armour than the T35 and the circular turret was subsequently abandoned in favour of a five-sided welded turret. As finalised, the design was standardised in June 1942 and designated M10 GMC. In order to increase production, use of the M4A3 chassis was also authorised and vehicles built on this chassis were designated M10A1 GMC. Most of these were retained in America for training or converted to prime movers, M35. Others were allocated to Lend-Lease shipments to Britain (see below). Grand Blanc Arsenal built 4,993 M10s between September 1942 and December 1943. Ford built 1,038 M10A1 between October 1942 and September 1943, and Grand Blanc built 675 M10A1, September-November 1943. 300 of the latter batch, however, were completed with new turrets as M36 (T71) GMC (qv).

Full-Track Prime Mover M35: M10A1 converted by removal of turret and fitting of air compressor and cables for towing 155mm and 240mm artillery pieces. Crew: 6; weight: 55,000lb.

British Service

A number of M10s and M10A1s were supplied to Britain in 1944 where they were designated "3in SP, Wolverine". These were issued for combat service to British units in Italy and France; most were converted from late 1944 by replacement of the 3in gun with the British 17pdr gun, producing a much more potent tank destroyer than the M10 in its original form. In its new guise the vehicle was designated "17pdr SP. Achilles Mk IC". M10A1s similarly converted were designated Achilles Mk IIC. The original mantlet was retained in this conversion. First in service in limited numbers in 21 Army Group in early 1945, the Achilles was used for many years post-war by the British. This was a most successful conversion.

Vehicles not converted were altered to gun towers by the removal of the turret and at least one of these was tested as an experimental mine plough.

379. M35 Prime Mover towing 240mm gun, France, early 1945.

380. British M10 gun tower converted to experimental mine plough, 1945.

381. T35 GMC was prototype for M10 series, utilising M4A2 hull.

382. T35E1 GMC showing revised sloped hull. Later it was given a five-sided turret. Prototype for M10 GMC.

SPECIFICATION

Designation: 105mm Gun Motor Carriage M10 or M10A1
Crew: 5 (commander, driver, gun crew (3))
Battle weight: 66,000lb
Dimensions: Length 19ft 7in; Height 8ft 1½in; Width 10ft; Track width 16½in; Track centres/tread 6ft 11in
Armament: Main: 1 × 3in gun M7 (1 × 17pdr OQF in Achilles)
Secondary: 1 × ·50 cal Browning MG (AA)
Armour thickness: Maximum 37mm
Minimum 12mm
Traverse: 360°. Elevation limits: +19° to −10°
Engine: Twin GMS 6—71 diesels (M10) Ford GAA V8 petrol (M10A)
Maximum speed: 30mph
Maximum cross-country speed: 20mph (approx)
Suspension type: Vertical volute
Road radius: 200 miles
Fording depth: 3ft
Vertical obstacle: 2ft
Trench crossing: 7ft 6in
Ammunition stowage: 54 rounds 3in
300 rounds ·50 cal MG

Special features/remarks: Standard M4 series chassis and motors but with entirely different well-sloped hull affording good armour protection. Retrospective modification was fitting of 2,500lb weight to turret rear to give better balance to turret. Late production vehicles had modified shape to rear of turret. British Achilles conversion with 17pdr gun was much superior in hitting power to original M10.

383. British 17pdr Achilles IIC conversion of M10A1, showing gun at maximum elevation.

384. Standard production M36. T71 GMC prototype vehicle was identical to this but lacked muzzle-brake.

IN October 1942 it was decided to investigate the possibility of adapting the 90mm AA gun as a high velocity anti-tank gun for mounting in American tanks and SP vehicles. In early 1943 a trial installation of a 90mm gun was made in the turret of the M10, but the gun proved too long and heavy for the turret which was, in any case, not entirely adequate for the 3in gun. In March 1943, therefore, work began on designing a new large turret to fit the M10 and take the 90mm gun. Tested at APG, the modified vehicle was very satisfactory and an initial "limited procurement" order of 500 vehicles was placed under the designation T71 GMC. In June 1944, the vehicle was standardised as the M36 GMC and entered service in NW Europe in late 1944 where it proved a most successful type able to knock out the heavy German Panther and Tiger tanks at long range. Some tank destroyer battalions notched up impressive scores with little loss to themselves using the M36. A priority programme to provide more M36 type vehicles to replace the less satisfactory M10 led to the following variants:

M36: Initial production type based on M10A1 chassis which was in turn based on M4A3 medium tank chassis. 300 produced by Grand Blanc April-July 1944 by completing M10A1 hulls as M36 vehicles with new guns and turrets. 413 produced by American Loco Co by converting existing M10A1, October-December 1944. 500 produced by Massey-Harris by converting existing M10A1, June-December 1944, 85 built by Montreal Loco Works, May-June 1945.

M36B1: An expedient design to meet increased demand for 90mm gun tank destroyers, this was produced by utilising the standard hull of the M4A3 medium tank fitted with the open-topped M36 type turret. 187 produced by Grand Blanc Arsenal, October-December 1944.

M36B2: Further expedient type utilising the M10 instead of the M10A1 hull. Several detail improvements including armoured covers for turret on some vehicles. 237 produced by converting existing M10 by American Loco, April-May 1945.

76mm Gun Motor Carriage T72: This was an interim design to overcome the shortcomings of the M10 which suffered from an unbalanced turret. The T72 was a M10A1 with a redesigned turret based on the T23 medium tank turret but with the top removed and thinner walls. There was a large rear "box" for a counterweight and the 76mm gun M1 was replaced by the 3in. However, it was decided to replace the M10 with the M18 Hellcat (qv) and the M36 so the T72 project was dropped.

SPECIFICATION

Designation: 90mm Gun Motor Carriage M36, M36B1, or M36B2
Crew: 5 (commander, driver, gun crew (3))
Battle weight: 62,000lb
Dimensions: Length 20ft 2in; Height 8ft 11in; Width 10ft; Track width $16\frac{1}{2}$in; Track centres/tread 6ft 11in
Armament: Main: 1 × 90mm gun M3
Secondary: 1 × ·50 cal Browning MG (AA)
Armour thickness: Maximum 50mm
Minimum 12mm
Traverse: 360°. Elevation limits: +20° to −10°
Engine: Twin GM 6–71 diesels (M36B2), Ford GAA V8 petrol (M36, M36B1)
Maximum speed: 30mph
Maximum cross-country speed: 18mph (approx)
Suspension type: Vertical volute
Road radius: 150 miles
Fording depth: 3ft
Vertical obstacle: 2ft
Trench crossing: 7ft 6in

Ammunition stowage: 47 rounds 90mm
1000 rounds ·50 cal MG
Special features/remarks: Overcame turret and gunpower deficiencies of the M10 series and proved a potent and impressive type in service. Many M36 vehicles were conversions from M10 series. Principal American tank destroyer type in final year of the war. Distinguished from M10 series by turret shape and long gun, occasionally seen with muzzle-brake removed.

385. M36B1 had the standard M4A3 medium tank chassis and hull and was easily distinguished from other M36 models. This vehicle lacks the usual muzzle-brake.

386. M36B2 utilised M10 hull and had detail changes including (usually) an armoured canopy over the open-topped turret.

387. T72 GMC had new turret and 76mm gun but never entered production.

MEDIUM TANK, M7

United States

388. Medium Tank, M7, production vehicle with 75mm gun.

FULL details of the development of this vehicle are given in the section of this book covering American light tanks. Originally designated Light Tank T7E2 it was reclassified and standardised as Medium Tank M7 in August 1942. Production orders were cancelled in February 1943, however, in the interests of standardisation on the M4 series as the one type of American medium tank. The M7 is shown here in its final production form for the sake of completeness. For T7E2 see plate 229.

389. Standard production M12 GMC. Recoil spade shown raised, and gun in travelling clamp.

THE M12 GMC was one of the earliest American SP weapons to be designed in World War II, but proved to be one of the last to see combat. Development of a 155mm gun on a modified M3 medium chassis was suggested in June 1941 by the Chief of Ordnance. A pilot model was completed by Rock Island Arsenal and tested at APG in February 1942, utilising the M1918 gun, a World War I weapon of French design used originally on a field carriage. Designated T6 GMC, the vehicle was initially rejected by Army Ground Forces who could not see any use for such a weapon. However, the Ordnance Department considered that an SP vehicle of this type was superior to a towed weapon of similar calibre. They therefore asked for 50 production vehicles in March 1942, but this was over-ruled by the Supply department pending a full test and assessment by the Artillery Board. The Board reported in support of the Ordnance Department's opinion and standardisation was commended in August 1942. Designated M12 GMC, the initial order was doubled instantly to 100 vehicles and production was completed in March 1943. Army Ground Forces were satisfied with this quantity and no more were ordered. Those built were used for artillery training in USA. However, in December 1943, with the invasion of Europe now in the planning stage, it was decided to overhaul 74 M12s for possible use in overseas theatres. This re-activation work was carried out by Baldwin between February-May 1944. They were sent to Europe, in June 1944 and were used for heavy bombardment work in several major actions, including the taking of Cologne.

VARIANT

Cargo Carrier M30: This was identical to the M12 except that the gun was omitted, as was the recoil spade at the back. The M30 was used as a "limber" vehicle for the M12, carrying the ammunition, battery stores, and gun crew. The M30 was issued on a scale of one for each M12 in service.

Both the M12 and M30 were based on the chassis of the M3 medium tank, but the engine was moved forward to the centre of the chassis to provide space for the gun mount. Driver and commander were seated in a separate front compartment with raised roof.

390. M12 GMC in action in Belgium, 1944, showing details of gun mount.

SPECIFICATION

Designation: 155mm Gun Motor Carriage M12, Cargo Carrier M30
Crew: 6 (commander, driver, gun crew (4) in M12, (other crewmen carried in M30))
Battle weight: 58,000lb (M12), 47,000lb (M30)
Dimensions: Length 22ft 1in (M30: 19ft 10in)
Height 8ft 10in (M30: 10ft)
Width: 8ft 9in
Track width 16½in
Track centres/tread 6ft 11in

Armament: Main: 1 × 155mm gun M1918M1
Secondary: 1 × ·50 cal Browning MG (AA) (M30 only)
Traverse: 14° each side. Elevation limits: +30° to −5°
Engine: Continental R–975 radial petrol 353hp
Maximum speed: 24mph
Maximum cross-country speed: 12mph
Suspension type: Vertical volute
Road radius: 140 miles
Fording depth: 3ft
Vertical obstacle: 2ft
Trench crossing: 7ft 6in
Ammunition stowage: 10 rounds 155mm (M12)
40 rounds 155mm (M30)
1,000 rounds ·50 cal MG (M30 only)
Special features/remarks: Based on M3 medium chassis but all except earliest production vehicles had M4 type bogies with trailing return rollers. Intended for long range bombardment in the field. M12 had recoil spade at rear; M30 had tailgate at rear.

391. Cargo Carrier M30 was limber vehicle for M12. It was identical except for omission of the gun.

GUN MOTOR CARRIAGE, M40 (T83) — United States

392. T83 GMC later standardised as M40 GMC.

DEVELOPMENT of an improved SP 155mm mount began in January 1944, following the decision to prepare the M12 (qv) for combat use in the forthcoming invasion of Europe. In December 1943 when it was decided to reactivate the limited numbers of M12s available, the Armored Force Board commended that for future armoured operations new designs of heavy self-propelled artillery should be prepared and units formed to operate this equipment for bombardment support to the armoured divisions. Army Ground Forces still did not consider this type of vehicle necessary, but the Ordnance Board supported the Armored Force and building of a pilot model was approved in March 1944.

At the end of 1943 it had been decided to concentrate a series of complementary AFV designs on the M4A3 chassis to form the so called "Medium Weight Combat Team" parallel to a "Light Weight Combat Team" on the M24 light tank chassis (qv). Within this "Medium Weight" concept, which was intended to rationalise maintenance and production, the new 155mm SP weapon fell. The new vehicle, designated T83 GMC, was thus designed on a slightly widened M4A3 medium tank chassis with HVSS. The design allowed for an interchangeable mount, in fact, with either the 155mm M1 gun or 8in M1 howitzer. With the latter armament, the vehicle was to be designated T89 HMC. To complete the family, a Cargo Carrier, T30 was designed which was simply the basic chassis less the weapon mount and modified to carry ammunition and crew members for either the T83 or T89. Following the invasion of Europe, an immediate "limited procurement" order for 304 T83s and 304 T30s was placed. Production was commenced by Pressed Steel in January 1945 and the T83 was standardised as the M40 GMC in March 1945. Requirements for M40 vehicles were subsequently raised to 600 and 311 were completed by the war's end. A few took part (with M12s) in the bombardment of Cologne, the first time they were used in action.

GUN MOTOR CARRIAGE, M12

VARIANTS

Cargo Carrier T30: Though 304 of these (see above) had been ordered in July 1944, they were cancelled in December 1944 in order to make the chassis available for T83 production, which by then was considered most urgent. Only a few T30s were built. A modified design, externally similar, had "universal" stowage racks to hold 105mm, 155mm, 8in, or 240mm ammunition, but this never entered production.

8in Howitzer Motor Carriage M43 (T89): Authority to build a pilot model of the T89 design was given in November 1944 and this (by conversion of a T83) was ready in early 1945. After trials at APG the T89 was approved in March 1945 and standardised in August 1945 as the M43. Production orders were, however, cut back at the cessation of hostilities and only 48 M43s were produced.

250mm (10in) Mortar Motor Carriage T94: The proved usefulness of the M12 GMC (qv) in bombardment work in NW Europe in late 1944-early 1945, led to the proposal to develop a vehicle specifically for siege work, basing it on the standard T83 SP chassis. Design of such a vehicle commenced in February 1945 and utilised the M40 (T83) chassis with a special mount for the T5E2 250mm mortar, heavier recoil spades, and a swivelling folding gantry for handling the ammunition and loading. Cessation of hostilities in August 1945 led to the cancellation of the project, though the pilot model was completed and tested in 1946. The mortar had an elevation range of 45–80° and a traverse of 15° either side.

BRITISH SERVICE

A small batch of M40s was delivered to Britain in post-war years and in British service this vehicle was designated "155mm SP, M40". They remained in service until about 1960.

SPECIFICATION

Designation: 155mm Gun Motor Carriage M40, 8in Gun Motor Carriage M43
Crew: 8 (commander, driver, gun-crew (6))
Battle weight: 80,020lb
Dimensions: Length 20ft 7in (29ft 9in over gun) Track width 23in
Height 8ft 9$\frac{1}{2}$in Track centres/tread 8ft 4$\frac{3}{4}$in
Width 10ft 4in
Armament: Main: 1 × 155mm gun M2
Secondary: —
Armour thickness: Maximum 12mm
Minimum 12mm
Traverse: 18° right and 18° left. Elevation limits: +45° to −5°
Engine: Continental R–975 radial
Maximum speed: 24mph
Maximum cross-country speed: 5–20mph
Road radius: 107 miles
Fording depth: 3ft
Vertical obstacle: 2ft 10in
Trench crossing: 7ft 8$\frac{1}{2}$in
Ammunition stowage: 20 rounds 155mm
Special features/remarks: Built on widened, lengthened M4A3 medium tank chassis as members of the "Medium Weight Combat Team". Mechanically similar to M4A3 except that Continental was used, moved to centre to give room for gun mount. Vision cupolas provided for driver and commander in separate compartment in hull front. Recoil spade and folding gun platform at rear. Otherwise layout similar to M12 GMC, but improved 155 mm gun in M40. Sometimes known (unofficially) as "Long Tom".

393. T30 Cargo Carrier.

394. T89 HMC, later standardised as the M43 HMC.

395. 250mm MMC, T94, sole prototype.

396. Crew of the first M40 GMC to fire on Cologne in 1945, pose in front of their vehicle.

397. Standard production M18 GMC, Hellcat, with M1A1C 76mm gun.

ORIGINS of the M18 GMC and its variants on the same chassis date back to December 1941 when the Ordnance Department recommended the development of a fast tank destroyer with 37mm gun on a chassis utilising Christie suspension and a Wright Continental R-975 engine. Two pilot models were authorised and the first was completed in mid 1942, designated T49 GMC. Changes in the original requirements included the substitution of the M1 57mm gun for the 37mm (due mainly to the need for more hitting power deduced from British experiences in the Western Desert), and the adoption of torsion bar, rather than Christie, suspension. The T49 was tested in July 1942 but the Tank Destroyer Command demanded a yet heavier gun and asked the Ordnance Department to complete the second pilot model with a 75mm M3 gun (as mounted in the M4 medium tank). The T49 project was thus cancelled in December 1942 in favour of the upgunned version, and the pilot model as completed was designated T67 GMC. It had a rounded, sloped, open-topped turret similar to that which had been produced for the T35 GMC (qv, M10 section). It was tested by the Armoured Vehicle Board and, due to its comparatively light weight of under 20 (short) tons, and powerful engine, it proved to have a lively performance, and was recommended for standardisation. In February 1943, however, Tank Destroyer Command again requested a more powerful gun, this time suggesting the 76mm M1 gun which was being developed for fitting to the M4 medium tank series. Six more pilot models were built all similar to the T67, but with the 76mm gun. These were designated T70 GMC. As a result of trials a few detail changes were made, including a modified (and simplified) shape for the hull front and a new turret which included a bustle for counter-weight and stowage box. On the whole, however, the T70 design was excellent needing little further modification for production, which started in July 1943 at the Buick factory. The vehicle was standardised in February 1944 as the M18 GMC, later being popularly called "Hellcat" though this was not official. 2,507 M18s were built, and production ceased in October 1944.

The M18 was one of the finest tank destroyers of any nation in World War II, and by virtue of its excellent power-to-weight ratio it was also the fastest tracked AFV to appear in that period. It also had the other virtues of a low silhouette, well-shaped armour protection, good reliability, and strong suspension, which made it a successful and popular vehicle with its crews. Hellcats were used by American tank destroyer battalions in both Italy and NW Europe, 1944-45, and the "hit and run" tactics possible with the superior performance allowed them to knock out a high total of enemy tanks with relatively little loss to themselves.

VARIANTS.

105mm Howitzer Motor Carriage T88: Following the proven success of the M18 in service, the Ordnance Department suggested in August 1944 the development of a similar vehicle mounting the 105mm T12 howitzer. A pilot model was completed in December 1944, identical in all respects to the M18 except for the gun and sighting arrangements. Designated T88, the project was cancelled while still under test, following the cessation of hostilities in August 1945.

Armoured Utility Vehicle T41 (M39): In June 1944 it was proposed to utilise the high speed/low silhouette characteristics of the M18 to produce a utility armoured vehicle capable of acting as a prime mover for the 3in M6 anti-tank gun (wheeled) or as a reconnaissance vehicle and troop carrier. Two M18s were modified by removal of the turret and revised internal layout. In prime mover form the vehicle was designated T41 and in reconnaissance form it was designated T41E1. Only difference was in the internal seating/stowage arrangements. A ring for a ·50 cal Browning AA machine gun was fitted at the front end of the fighting compartment. Details as M18, except: Weight: 35,000lb; crew: 2 plus 7 gun crew or infantrymen; height: 5ft 11in (excluding gun mount). Ordered under "limited procurement" in June 1944, it was standardised as the M39 in early 1945.

398. T49 37mm GMC shown under test.

399. One of the T70 pilot models; identical to the T67 GMC except for the 76mm M1 gun.

76mm Gun Motor Carriage T86, T86E1, and T87 (Amphibious): Combat experience in the Pacific led to several experiments and projects to give amphibious capability to US AFVs. In January 1944 two pilot models of an amphibious version of the M18 were proposed, retaining all original features but with a large lightweight flotation hull. The first pilot, T86 GMC, was propelled by its tracks in the water, while the second was propelled by two 26in propellers driven from a power take-off in the engine. This vehicle was designated T86E1. The best of these two propulsion methods was to be used in the T87. This proved to be the track type, and an improved track was incorporated in the T87 along with detail changes to the hull. The T87 appeared in December 1944 and was still undergoing trials at the cessation of hostilities, after which it was cancelled. The T87 had the same 105mm howitzer as the T88 while the T86/T86E1 had the 76mm gun of the M18. The T87 had a slightly shorter hull. All these prototypes proved satisfactory on test with a good performance in surf. However, forward vision was generally poor due to the hull shape. In addition it was found necessary to add cable-controlled rudders at the hull rear to assist steering.

400. T88 105mm HMC; sole pilot model.

401. Standard production T41 (M39) Armoured Utility Vehicle.

SPECIFICATION

Designation: 76mm Gun Motor Carriage M18
Crew: 5 (commander, driver, gunners (3))
Battle weight: 40,000lb
Dimensions: Length 21ft 10in (17ft 4in excluding gun)
Height 8ft 5in (including AA mount)
Width 9ft 9in
Track width 14·4in
Track centres/tread 10ft $10\frac{1}{4}$in
Armament: Main: 1 × 76mm gun M1A1, M1A1C, or M1A2
Secondary: 1 × ·50 cal Browning MA (AA)
Armour thickness: Maximum 12mm
Minimum 7mm
Traverse: 360°. Elevation limits: $+19\frac{1}{2}^\circ$ to -10°
Engine: Continental R-975 air cooled petrol
400 hp
Maximum speed: 45-50mph
Maximum cross-country speed: 20mph (approx)
Suspension type: Torsion bar
Road radius: 150 miles
Fording depth: 4ft
Vertical obstacle: 3ft
Trench crossing: 6ft 2in
Ammunition stowage: 45 rounds 76mm
800 rounds ·50 cal
(M41: 42 rounds 3in, in prime mover role only)
Special features/remarks: Highly successful type which originated the torsion bar suspension adopted for later vehicles from the M24 light tank onwards. Early production M18s had M1 gun or M1A1 gun without muzzle brake. Compensating drive sprocket to take up tension in track over rough terrain. All welded construction with open topped turret. All variants were mechanically identical.

402. T86 GMC (Amphibious) showing new flotation hull on M18 chassis.

403. T87 GMC (Amphibious) with 105mm howitzer and modified hull.

404. The T20E3 was the second model completed in the T20 series and had torsion bar suspension.

ONCE the M4 series had reached production status, consideration was immediately given to its successor. On May 25, 1942, the Ordnance Department received confirmation from the Supply Services that it could go ahead with designing and procuring a pilot model for an improved medium tank, provisionally designated M4X. Broad requirements called for a 32 (short) ton vehicle with automatic 75mm gun, 4in of front armour and a top speed of 25mph. A wooden mock-up was accordingly built by Fisher, one of the medium tank producers. By September 1942 it had been agreed by the Ordnance Department, after consulting the Armored Force Board, to build three pilot models of 30 tons maximum weight, each with different armament and interchangeable turrets. The first one, T20, was to have a 76mm gun and HVSS, the second, T20E1, was to have HVSS and a 75mm automatic gun, while the third, T20E2, was to have a 3in gun and torsion bar suspension. Each was to be powered by the new Ford GAN V-8 tank engine and have a torque converter and Hydra-matic transmission. The T20 was built by Fisher and completed in June 1943. The T20E1 was cancelled but its turret was used in the T22E1 (qv).

The T20E2 was completed by Fisher as the T20E3 with 76mm, instead of 3in gun.

Both the T20 and T20E3 were tested but the transmission system adopted gave much trouble with oil leaks and overheating. Development of these vehicles ceased at the end of 1944, by which time developments had proceeded to much later types. Experience and information gathered from these vehicles under test was useful, however, for development of later T20 range tanks. Maximum armour thickness of these T20 tanks was 62mm at the front, they had a 47° sloped hull front, were all welded with a cast turret, and featured several standard fittings from the M4 series. In the T20 series, drive was to the rear.

SPECIFICATION:

Designation: Medium Tank T20, T20E3
Crew: 5 (commander, driver, co-driver, gunner, loader)
Battle weight: 65,758lb (T20), 67,500lb (T20E3)

Dimensions: Length: 18ft 10in (excluding gun) Track width $16\frac{1}{2}$in (T20) 18in (T20E3)
Height 8ft Track centres/tread: —
Width 9ft 10in
Armament: Main: 1 × 76mm gun M1
Secondary: 2 × ·30 cal Browning MG
1 × ·50 cal Browning MG (AA)
Armour thickness: Maximum 62mm
Minimum 12mm
Traverse: 360°. Elevation limits: +25° to −10°
Engine: Ford GAN V-8 petrol
470hp
Maximum speed: 25mph (35mph—T20E3)
Maximum cross-country speed: —
Suspension type: HVSS (T20)
Torsion bar (T20E3)
Road radius: 100 miles
Fording depth: 4ft 8in
Vertical obstacle: —
Trench crossing: —
Ammunition stowage: 70 rounds 76mm
6000 rounds ·30 cal
Special features/remarks: Virtually an improved M4 with lower silhouette. Most characteristics similar to M4 except for transmission features and hull shape. Prototypes only.

405. Medium Tank T20 showing early form of horizontal volute suspension and other features from the M4 series.

406. Medium Tank T22, first pilot model.

407. T22E1 with 75mm automatic gun was the first pilot T22 converted with the turret designed for the T20E1.

THE T22 was a development of the T20 series (qv) initiated in October 1942. Chrysler were asked to build two pilot tanks identical in all respects to the T20 except for the transmission which was to be of the same five-speed mechanical type as used in the M4 medium tank. Trouble was experienced with the transmission and rear drive, however, on tests, and work on the T22 project was formally cancelled in December 1944. Both vehicles had been completed in June 1943.

Subsequently the first pilot model was converted to take the special turret with 75mm automatic gun which United Shoe Machinery Corp had built for the projected T20E1 (qv). In its new guise, the T22 was redesignated T22E1. The turret was virtually a lengthened M4 type, the gun was the standard M3 weapon, and the mount was the standard M34 type. Main feature was the automatic hydraulic loader with two magazines, one for HE and one for AP ammunition, selected remotely as required by the commander who was located in the left rear of the turret. The only other turret occupant was the gunner, and there was no loader. Tests at APG gave a maximum rate of fire of 20 rounds per minute, but the magazines and loading mechanism were unreliable.

By this time, however, there was a requirement for a heavier calibre gun, and the project was cancelled in December 1944. Apart from the transmission (and the turret in the T22E1) data and characteristics for the T22 series were the same as for the T20. T22E2 was the projected equivalent of the T20E2 with 3in gun. It was cancelled in the design stage.

408. Standard production T23 medium tank.

DEVELOPMENT of the T23 was authorised at the same time as that of the T22. Hull, armament, and general external layout were similar to the T22, but vertical volute suspension and tracks as used on the M4 series were to be fitted and electric transmission was specified, the drive units being built by General Electric. As with the T22, three pilot models were asked for, the T23 with 76mm gun, the T23E1 with 75mm automatic gun, and the T23E2 with 3in gun. However, as in the T20 series, the projects for vehicles with 75mm gun and 3in gun were cancelled before completion. The T23 pilot model was, in fact, the first of the T20 type tanks to be completed, finished by Detroit Arsenal in January 1943 and under test before either the T20 or T22. A second pilot model was ready by March 1943. The pilot models were tested at Fort Knox and proved to be very manoeuvrable. In May 1943 a "limited procurement" order of 250 vehicles was commended, subject to detail improvements. These vehicles were built by Detroit Arsenal between November 1943 and December 1944, differing from the pilot models in having an all-round vision cupola for the commander, a rotating hatch for the loader, an improved gun mount (the T80), and an improved 76mm gun, the M1A1. Though used in limited numbers in America, the T23 was never standardised and never generally issued or used in combat.

Main reason for this was that Army Ground Forces were already quite satisfied with the M4 medium tank while the Armored Force Board considered some T23 features unsatisfactory, in particular poor weight distribution, excessive ground pressure, and a mode of transmission which was untried in the long term and possibly suspect. They requested ten T23s for further trials in an attempt to overcome these shortcomings. First of these ordered was designated T23E3 and was to have torsion bar suspension and 19in tracks, while the second, T23E4 was to have HVSS and wide tracks. This latter vehicle was subsequently cancelled however.

The T23E3, completed by Chrysler at Detroit Arsenal in August 1944, had a turret taken from a production T23 and torsion bar suspension taken from the T25E1 (qv). All other features were the same as the T23 but the turret basket was eliminated to give increased ammunition stowage and the electric transmission was fully waterproofed, a retrospective modification also featured in late production T23s. On the basis of experience with the T20E3, which had by this time been completed with torsion bar suspension, the Ordnance Department requested in July 1943 that the T23E3 be standardised as the Medium Tank M27 and the T20E3 be standardised as the M27B1 both for immediate production in view of the fact that the M4 medium tanks would be seriously obsolescent by 1944. This was rejected by Army Service Forces and no further progress was made with standardising the T23 series.

In the event, however, this led to numerous improvements being made in the M4 series for introduction from late 1943 onwards and many of the features tried or developed in the T20-T23 tanks were incorporated into M4 vehicles, in particular the complete T23 type turret and 76mm gun, HVSS, and the simplified 47° hull front. Thus, while the T20-T23 series vehicles did not see general service or combat, they played a most important part in US tank development in the late war period leading to improvements in the M4 design and, as developed into the T25 and T26 (qv), leading to the evolution of the M26 heavy tank.

409. Medium Tank T23, second pilot model with prototype turret fittings and narrow T79 gun mount.

MEDIUM TANK, T23

SPECIFICATION:

Designation: Medium Tank T23, T23E3.
Crew: 5 (commander, driver, co-driver, gunner, loader)
Battle weight: 73,500lb (T23), 75,000lb (T23E3)
Dimensions: Length 19ft 8⅞in (T23) Track width 16½in (T23)
19ft 2⅛in (T23E3) 19in (T23E3)
Height 8ft 4in (T23) Track centres/treat: 8ft 2¼in
8ft 4⅝in (T23E3)
Width 9ft 10½in (T23)
10ft 4in (T23E3)
Armament: Main: 1 × 76mm gun M1A1
Secondary: 2 × ·30 cal Browning MG
1 × ·50 cal Browning MG (AA)
Armour thickness: Maximum 87mm
Minimum 12mm
Traverse: 360°. Elevations limits: +25° to −10°
Engine: Ford GAN V-8 gasoline with electric transmission 470hp
Maximum speed: 35mph
Maximum cross-country speed: 20mph
Suspension type: Vertical volute (T23)
Torsion Bar (T23E3)
Road radius: 100 miles
Fording depth: 4ft 8 in
Vertical obstacle: 2ft

410. Medium Tank T23E3.

Trench crossing: 8ft
Ammunition stowage: 84 rounds 76mm (T23E3)
64 rounds 76mm (T23)
Special features/remarks: Weights quoted are net; add about 4,000lb for approximate combat weight. Other remarks as T20 and T22 series. T23 was only design of T20-T23 range to achieve production (but not combat) status. T23E3 had 20% lower ground pressure than the T23 and overcame the major shortcoming—high ground pressure—of the basic T23 design.

MEDIUM TANK, T25 SERIES

United States

411. Medium Tank T25, first pilot model.

IN September 1942 when the T20 design was drawn up, the Ordnance Department suggested that a 90mm gun be developed for future mounting in tanks of this series. By March 1943, such a weapon had been produced in prototype form and mounted for tests in one of the T23 pilot models, these being the first vehicles completed in the T20-T23 range.

In April 1943, Army Service Forces gave approval for the procurement of 50 trials vehicles, based on the T23 but mounting 90mm guns. Of these, 40 were to have the same basic armour characteristics as the T23 and would be designated T25, while the other 10 were to have increased armour protection to provide a "heavy" medium tank of comparable performance and immunity to the German Tiger tank which had just then made its appearance in action in Tunisia. The heavier design was to be designated T26.

Detroit Arsenal built two pilot models of the T25, completing them in January and April 1944 respectively. They had HVSS and 23in tracks, and Ford engine and electric drive as in the T23. The hull was reinforced with a ribbed hull top to support the massive cast turret and had internal modifications to allow the stowage of 90mm ammunition. The T25 pilot models were tested at Fort Knox, but by this time interest had switched to the T25E1 design which came about when design studies for the T25 showed that this vehicle with its electric transmission would weigh more than 40 (short) tons.

To reduce weight as much as possible it was decided to drop the electric transmission and revert to the Hydramatic type transmission with torque converter as had been featured in the T20 medium tank. Accordingly the order for 40 T25s was switched to this modified design, designated T25E1, and Grand Blanc Arsenal (Fisher), the contractors, completed the first in January 1944 and finished the production run the following May. By this time, however, with the invasion of Europe approaching, attention had switched to the more heavily armoured T26 series (qv), and the 40 T25E1 vehicles were used solely for tests and development work. The T25E1 differed from the T25 in having torsion bar suspension and a modified hull as well as different transmission. Suspension from a T25E1 was used for the T23E3 (qv).

SPECIFICATION:

Designation: Medium Tank T25E1
Crew: 5 (commander, driver, co-driver, gunner, loader)
Battle weight: 77,590lb
Dimensions: Length 22ft 11¼in; Height 9ft 1⅜in; Width 10ft 4in; Track width 23in; Track centres/tread 9ft
Armament: Main: 1 × 90mm gun T7 (M3)
Secondary: 2 × ·30 cal Browning MG; 1 × ·50 cal Browning MG (AA)
Armour thickness: Maximum 87mm; Minimum 12mm
Traverse: 360°. Elevation limits: +20° to −10°
Engine: 1 × Ford GAF V-8 petrol 470hp
Maximum speed: 30mph (approx)
Maximum cross-country speed: 20mph
Suspension type: Torsion bar
Road radius: 100 miles
Fording depth: 4ft 8in
Vertical obstacle: 2ft
Trench crossing: 7ft 6in
Ammunition stowage: 48 rounds 90mm; 300 rounds ·50 cal; 5,00 rounds·30 cal
Special features/remarks: T25 had similar characteristics but weighed about 81,000lb and had HVSS with 23in tracks. It had the same engine, but with electric drive.

412. Standard production T25E1. Note modified hull front compared to T25.

413. Medium Tank T26E1, first pilot model.

CONCURRENTLY with the design of the T25 (qv), designs were drawn up for a more heavily armoured version, designated T26. This would have had electric transmission, as in the T25. Development studies showed that both the T25 and T26 with this transmission would be excessively heavy, the T26 weighing more than 45 (short) tons. In view of this, the T25 was redesigned with Torquematic transmission (Hydra-matic transmission with a torque converter) and the same changes were, of course, incorporated in the T26 design. Only the T26 pilot model was therefore built and the 10 scheduled production vehicles for test purposes were all completed as T26E1s by Grand Blanc Arsenal in March-June 1944. The T26E1 was similar in most characteristics to the T25E1 but had wider tracks (24in), increased overall width (11ft 2in), a shorter hull (22ft 4¾in), increased armour maximum (100mm), and a correspondingly increased weight (86,500lb). Other details were the same as those of the T25E1.

Meanwhile in September 1943, the Ordnance Department had urged immediate production of 500 T25E1s and 500 T26E1s for delivery in 1944, but this was opposed by both the Armored Force Board, who would have preferred the 90mm gun mounted in the M4 medium tank, and by the commander of Army Ground Forces who did not consider a 90mm gun desirable in a tank since it would encourage tank units to stalk enemy tanks, a role assigned to tank destroyers in the then-current armour doctrine of the US Army. Army Ground Forces, instead, requested in April 7,000 T25E1s with 75mm guns and 1,000 T26E1s armed with a 76mm gun. This clearly impractical request—which would have involved further development work and, in any case, duplicated existing types with smaller calibre guns—was not resolved until June 1944 when all the T26E1 development vehicles had been completed. On June 1st, a statement came from European Theatre of Operations that they required no new vehicles with 75mm or 76mm guns in 1945; instead they wanted tanks with 90mm and 105mm weapons in the ratio 1:4. Their request was upheld by the Army Staff and the T26E1 underwent its trials programme as planned. At the end of June 1944 the T26E1 was reclassified Heavy Tank T26E1 and its development history as prototype for the M26 Pershing is accordingly continued in the American heavy tank section of this book.

414. Rear detail view of Medium Tank T26E1.

415. Standard production Heavy Tank M6A1 (T1E3).

AT the end of May 1940, the Chief of Infantry, then still responsible for the US Army's tanks, formulated suggestions for future tank types in the light of events in Europe where Germany had just over-run France largely by the brilliant use of armoured divisions. Some PzKw IVs with 75mm guns had appeared in action, rendering all American types with the 37mm gun technically obsolete. The suggestions were passed to the Ordnance Department early in June and called for two types of new vehicle. One was an improved medium tank of the M2A1 type but with 75mm gun, and progress with this idea led within the next two months to development of the M3 medium tank (qv). The second new type requested was a heavy tank in the 80 ton class which, as first envisaged, would have two turrets each with a 75mm gun and partial traverse, and two smaller turrets with full traverse, each with a 37mm gun, in addition to a 20mm and ·30 cal machine gun. Armour minimum was to be 75mm. These characteristics were subsequently modified to call for a hull-mounted gun of larger calibre than 75mm, with a turret-mounted gun of 37-50mm calibre, plus eight machine guns—virtually a scaled-up M3 medium tank.

By late October 1940, however, when the Ordnance Department had finalised plans for what was by then designated Heavy Tank T1, the specification had been revised to give a vehicle of about 50 (short) tons, with 75mm armour, a 3in and 37mm gun mounted coaxially in a fully traversing turret, four machine guns, a Wright 925HP engine, Hydra-matic transmission, and a top speed of 25mph. In February 1941 approval was given to build four pilot vehicles and plans were made to produce the new vehicle at the rate of 100 a month. The four pilot vehicles were to test alternative forms of transmission and hull form, to select the best for the production models. The T1E1 was to have a cast hull, and electric transmission, the T1E2 was to have a cast hull and torque converter, the T1E3 was to have a welded hull and torque converter, while the T1E4 was to have welded hull and two twin diesel motors with two torque converters. This latter was subsequently cancelled since the diesel installation would have involved protracted development time.

A contract had been placed with Baldwin for the pilot models, and the first to be completed was the T1E2 in December 1941, the day after the attack on Pearl Harbor and America's declaration of war. Trials with this vehicle at APG showed the need for modifications to the brakes and the cooling system, and when this work was completed satisfactorily in April 1942, the T1E2 was standardised as the M6. Meanwhile the T1E3 had been completed and tested and this was standardised as the M6A1, the only external difference being the welded hull of the latter. The T1E1 was last to be completed and was not ready for testing (at Fort Knox) until June 1943. This latter variant was never standardised, though it became semi-officially known as the M6A2 and was generally referred to as such.

When the first two pilot models were standardised in the spring of 1942, the Ordnance Department more than doubled their original production target to 250 vehicles a month, bringing in Grand Blanc Arsenal (Fisher) as a second contractor in addition to Baldwin. These were the crisis days for America, however, when the President's Victory Program called for a huge increase in the army with a big concentration on tank production. By September 1942, the new Army Supply Program was introduced which cut back tank output in favour of more aircraft. The M6 was a prominent victim of this, a target for 5,000 being slashed to only 115. Meanwhile the Armored Force had been testing the pilot models and reported on December 7, 1942, that they considered the M6 unsatisfactory, being too heavy, under-gunned, poorly shaped, and requiring improvements to the transmission. Because of these deficiencies and consequent tactical limitations they could see no requirement for heavy tanks of this type. In March 1943, therefore, the Ordnance Department cut back production requirements to only 40 vehicles, 8 M6, 12 M6A1, and 20 M6A2. All were built by Baldwin between

HEAVY TANK, M6

416. Standard production M6A2 (T1E1); note cast hull compared with welded hull in M6A1, plate 415.

418. T1E1 fitted experimentally with 90mm gun T7.

417. The T1E2 pilot model which was standardised as the M6; cupola and rear facing AA machine gun in turret were dropped from production vehicles.

419. M26A2E1, first pilot model, with 105mm gun and new cast turret, 1945.

November 1942 and February 1944.

These M6 series vehicles were thus never issued for combat but they were used for trials and experimental work. The T1E1 electric transmission system was used in the T23 medium tanks (qv), and in early 1944, one M6 was modified for trials with a 90mm gun from the T26E1 medium tank (qv), though the project was cancelled in March 1944. In July 1944 when there was now an urgent requirement for heavy tanks in ETO, one M6A2 was modified with a new heavy turret and 105mm gun for possible use in this role. It was planned to convert 15 M6A2 vehicles for shipment to Europe with this 105mm gun, but the idea was not adopted and the project was dropped. Thus modified, the test vehicle was designated M6A2E1. In December 1944 the M6 series was declared obsolete.

The M6A2E1, plus a second vehicle similarly converted, was used in mid 1945 to test the gun, mount, fittings, and internal installations for the T29 heavy tank (qv), then being developed, which was to mount the 105mm gun.

When designed, the M6 was the heaviest and most powerfully armed tank in the world, though it was soon outdated by other vehicles which took this distinction. US armoured doctrine followed the German example in the early war years and concentrated on the fast medium tank, so there was little enthusiasm for heavy tanks in the Armored Force in 1942. Ironically, the Germans had switched their emphasis to heavy tanks like the Tiger and Panther by the time the American and German forces clashed in 1944, by which time the M6 had faded from the scene and a new heavy tank, the M26, was being developed from the T20 range of medium tanks. Some of the features first tried in the M6, like Torquematic transmission and rear drive, were, however, perpetuated in the T20 range of medium/heavy tanks.

SPECIFICATION:

Designation: Heavy Tanks, M6, M6A1, T1E1 (M6A2)
Crew: 6 (commander, driver, co-driver, gunner, loaders (2))
Battle weight: 126,500lb
Dimensions: Length 27ft 8in; 24ft 9in (hull only)
Track width $25\frac{3}{4}$in
Track centres/tread 7ft 9in
Height 10ft 7in
Width 10ft $2\frac{1}{2}$in
Armament: Main: 1 × 3in gun M7 and 1 × 37mm gun M6 (co-axial mount)
Secondary: 2 × ·50 cal Browning MG (bowl)
2 × ·30 cal Browning MG, 1 × ·50 cal MG (AA)
Armour thickness: Maximum 100mm
Minimum 25mm
Traverse: 360°. Elevation limits: +30° to −10°
Engine: Wright G-200 9 cylinder radial, air cooled 800hp (electric transmission in T1E1)
Maximum speed: 22mph
Maximum cross-country speed: —
Suspension type: HVSS with double bogies of 4 wheels each and twin tracks each side
Road radius: 100 miles (approx)
Fording depth: 4ft
Vertical obstacle: 3ft
Trench crossing: 11ft
Ammunition stowage: 75 rounds 3in
202 rounds 37mm
5,700 rounds ·50 cal, 7,500 rounds ·30 cal
Special features/remarks: M6A1 had welded hull, others cast hull. Production vehicles differed in detail from prototypes. Prototypes also had machine gun cupola as fitted to M3 medium tanks, plus a rear firing AA machine gun, both features being eliminated in production vehicles. Tracks were steel with rubber inserts on inner sides. Rear drive with front idler and auxiliary "jockey" wheel just behind it. Twin tracks, twin bogies, and HVSS.

420. First of the two T14 pilot models.

ANGLO-AMERICAN co-operation in the exchange of ideas and information relating to tank policy and procurement started in June 1940 when the British Tank Mission arrived in the USA to obtain tanks for the British Army. The M3 Grant, the M3 medium modified to meet British requirements, was an early result of this co-operation. Major-General Charles M. Wesson, Chief of Ordnance (head of Ordnance Department), led an American mission to Britain in September 1941 to learn first-hand about British tank combat experience and hear British views on American equipment and British requirements for the future. Among ideas discussed was the case for a vehicle with large calibre gun and heavy armour for which the British at that time considered there was a need, following recent experience against the Germans in the Western Desert fighting. (Partly as a result of this thinking and partly because of disappointment with the early models of the Churchill infantry tank, the British proposed a heavier version of the A27L (Centaur), the A28, and variations on this project subsequently led to the A33 assault tank as a possible replacement for the Churchill. Further details on this line of policy are given in the A33 history in the British section of this book.)

British views at the September 1941 meeting appealed to the US Ordnance Department who at that time were working on the M6 heavy tank (qv), and were advocating the need for heavy tanks in the US Army. In December 1941, the Ordnance Department commenced design studies for a heavy assault tank which broadly met the stated British requirements and incorporated some features similar to the T1 (M6) heavy tank and the new M4 medium tank, with the intention that it should utilise as many M4 components as possible. It was to have a 75mm gun M3 with alternative provision for a British 6pdr gun, and armour maximum was to be 75-100mm. The new Ford tank engine was to be used with the possibility of an enlarged engine of the same type later. In March 1942 a new British Tank Mission went to the United States mainly to settle procurement problems, but included in business discussed was the possibility of the American assault tank design, now designated T14, being built in America for Britain. An agreement for 8,500 vehicles was concluded with the Ordnance Department, and work on detailed design commenced. Two pilot models were completed in 1943, but trials showed that modifications were necessary to the track and suspension. One of the T14 pilot models was sent to Britain for tests and evaluation in 1944, but by this time British tank policy had changed in favour of cruiser tanks with large calibre guns, and the Churchill had proved itself and been retained as the heavy infantry tank. Thus the British interest in the T14 had waned and the Ordnance Department dropped the complete project in December 1944 and the vehicle never entered production.

As completed the T14 utilised a complete transmission system identical to the M4 and, the standard Ford tank engine. The final drive was, however, geared down to give a slower maximum speed. Armament matched that of the M4 and the tracks and suspension units were adapted from the type used in the M6.

SPECIFICATION:

Designation: Assault Tank T14
Crew: 5 (commander, driver, co-driver, gunner, loader)
Battle weight: 84,000lb
Dimensions: Length 20ft 4in; Height 9ft 1in; Width 10ft 3in; Track width 25¾in; Track centres/tread 7ft 9in
Armament: Main: 1 × 75mm gun M3
Secondary: 2 × ·30 cal Browning MG
1 × ·50 cal Browning MG (AA)
Armour thickness: Maximum 133mm
Minimum 19mm
Engine: Ford GAZ V-8 petrol 520hp.
520hp
Maximum speed: 22mph
Maximum cross-country speed: —
Suspension type: HVSS type and tracks as on M6
Road radius: 100 miles
Fording depth: 3ft
Vertical obstacle: 2ft 1in
Trench crossing: 9ft
Ammunition stowage: 50 rounds 75 mm
9,000 rounds ·30 cal
Special features/remarks: Heavy skirt side armour and cast hull and strong "family resemblance" to M4 medium series. As finally built, the T14 had provision for mounting a 76mm or 105mm gun in place of the 75mm weapon. Fitted with British radio equipment (No 19 set).

421. Standard production Heavy Tank T26E3 (M26), Pershing.

IN June 1944, as related previously (see T26 medium tank series), the T26E1 was redesignated Heavy Tank T26E1. Extensive trials were carried out with the ten T26E1 pilot models by the Ordnance Department and numerous detail modifications were made for incorporation in production vehicles. These included improvements to the transmission and the engine cooling, revised electrical system, removal of the turret cage to increase ammunition stowage, better engine access, and larger air cleaners. In August 1944, the Ordnance Department recommended that the T26E1 be standardised and placed into production. Opposition from the user arms was still strong, however, and Army Ground Forces disagreed and stated that the vehicle could not be standardised until the Armored Force Board had also tested and approved the production modifications. Earlier, in July, Army Ground Forces had tried another delaying move by requesting that the T26E1 be redesigned with the 76mm gun, a retrograde idea ignored by the Ordnance Department. It was not until December 1944 that the T26E1 was approved for "limited procurement" and production vehicles, with the various modifications earlier suggested, were designated T26E3 to distinguish them from the pilot models.

Production of the first 20 T26E3s had begun in November and the Ordnance Department proposed early in December that these be shipped straight to Europe for combat testing. Once again Army Ground Forces was opposed to the idea and asked that they first go to the Armored Force for testing and "certification of battleworthiness". This would have wasted yet another month. However, within a week of this exchange, two German panzer armies savagely hammered the US 1st Army in the lightning Ardennes Offensive, December 16, 1944. Among other things, this reverse spotlighted the inadequacies of the M4 medium with its relatively light armour and 76mm gun; undoubtedly this was a factor which caused the American General Staff to intervene in the T26 affair on December 22, and order immediate shipment of available T26E3s to Europe without further testing.

The first 20 T26E3s were shipped to Europe in January 1945 and at the beginning of February these were issued for service to the 3rd and 9th Armored Divisions. The Ardennes Offensive had indeed vindicated the Ordnance Department's persistent attempts to get a tank with a 90mm gun into service. In January 1945 Army Ground Forces had no hesitation in agreeing that the T26E3 be considered battleworthy, and the cry for more vehicles came from ETO where tank crews were favourably impressed with the new tank which was nearly a match for the Tiger in a straight shooting match, and very much more mobile. Full production of the T26E3 was ordered in January 1945 and it was built by Grand Blanc Arsenal (1,190, November 1944-June 1945) and later Detroit Arsenal (246, March-June 1945). In March 1945 the T26E3 was standardised as the Heavy Tank M26 and it was named "General Pershing", usually shortened to "Pershing". Later in 1945, the M26 saw action in the Pacific, being used in the taking of Okinawa.

PRODUCTION VARIANTS

M26: See main text. In "limited procurement" status initially as T26E3.

M26E1: The M26 with its 90mm M3 gun still had inferior hitting power to the German Tiger with its 88mm gun. To improve the performance of the M26 a new longer gun was developed, the T54, which fired fixed ammunition using a larger cartridge case. This weapon had a concentric recoil mechanism. Two pilot models with the new gun were approved in May 1945, but no production order followed due to the cessation of hostilities.

T26E4: This was the development model for the M26E1 and was essentially the M26 with a longer 90mm gun T15E2 which fired separate ammunition with a heavier charge in an attempt to match the hitting power of the German 88mm gun. Due to the longer barrel it was necessary to modify the gun cradle and elevating mechanism, and add a counterweight in the turret. A total of 25 vehicles of this type were produced from March 1945, classified "limited procurement", though 1,000 had been authorised prior to the cessation of hostilities.

T26E5: This was a heavy assault version of the M26 with thicker frontal armour on hull, turret, and mantlet. In its final form it had a heavier turret with 11in of armour on the mantlet, and 6in of frontal armour on the hull. Weight was 51 (short) tons. The track was permanently fitted with grousers. Classified "limited procurement", 27 were built, commencing June 1945, all at Detroit Arsenal. The added weight of the T26E5 had an adverse effect on performance and this vehicle was not considered entirely satisfactory.

M45 (T26E2): To meet the requirement from ETO for tanks with the 105mm howitzer, a weapon of this calibre was tested in a T23 medium tank (qv). In June 1944 this was dropped in favour of the same weapon in the T26E1. Designated T26E2, the vehicle had a redesigned turret to take the howitzer, and modified ammunition stowage. A heavier turret front was fitted to maintain equilibrium of the turret with the lighter weapon. In July 1945 the T26E2 was classified a "limited procurement" type and redesignated M45. Only a small number were produced. Weight: 92,500 lb; ammunition: 74 rounds 105mm; elevation: $-10°$ to $+35°$. Other details as M26.

M28A1: This was the M26 with the improved 90mm M3A1 gun. Developed postwar. In May 1946, all T26/M26 series vehicles were reclassified as medium tanks.

8in Howitzer Motor Carriage T84: In early 1944 development had started of the 155mm GMC T83 on the M4A3 chassis, the vehicle later standardised as the M40 GMC (qv). In April 1944, it was suggested by the Artillery Board that the 8in howitzer could be mounted on the same chassis. Army Service Forces approved the development of a self-propelled 8in howitzer, but asked that it be based on the new T26E1 chassis. Two pilot models were built, but were subject to much delay due to the lack of availability of T26E1 chassis. Trials were under way with the pilot models when the war ended and the project was cancelled. As a result of the delays with this vehicle, the M4A3 chassis was, after all, used as a basis for the 8in HMC, becoming the T89 (M43) as described in the M40 section.

422. M26E1 with T54 gun. Note counterweight on turret rear.

423. T26E5 assault tank. Note grousers on track.

424. M45 (T26E2) with 105mm howitzer.

425. T26E4 with T15E2 gun. Note extra barrel length and counterweight on turret rear.

Cargo Carrier T31: This was the ammunition carrier designed to accompany the T84 HMC. It incorporated a cargo compartment in place of the howitzer mount. Work on the prototype started in April 1944 and tests were under way when the war ended.

426. T84 8in Howitzer Motor Carriage.

427. T31 Cargo Carrier.

SPECIFICATION:

Designation: Heavy Tank M26 (T26E3), Pershing
Crew: 5 (commander, driver, co-driver, gunner, loader)
Battle weight: 92,000lb
Dimensions: Length 28ft 10in
21ft 2in (hull only)
Height 9ft 1in
Width 11ft 6in
Track width 24in
Track centres/tread 9ft 2in
Armament: Main: 1 × 90mm gun M3
Secondary: 2 × ·30 cal Browning MG
1 × ·50 cal Browning MG (AA)
Armour thickness: Maximum 102mm
Minimum 13mm
Traverse: 360°. Elevation limits: +20° to −10°
Engine: Ford GAF V-8 petrol
500hp
Maximum speed: 20mph
Maximum cross-country speed: 5·2mph
Suspension type: Torsion bar
Road radius: 92 miles
Fording depth: 4ft
Vertical obstacle: 3ft 10in
Trench crossing: 8ft 6in
Ammunition stowage: 70 rounds 90mm
5,000 rounds ·30 cal
550 rounds ·50 cal
Special features/remarks: Turret included vision cupola for commander and roof hatch for loader. Two escape doors were fitted in the hull floor. Rear drive and torsion bar suspension were inherited from previous vehicles in the T20-T25 range, and the vehicle represented the culmination of development of this series in World War II. Further developments in postwar years have led to the standard M60 battle tank of the US Army in the sixties.

BRITISH SERVICE

In 1945 a small number of M26 tanks was supplied to Britain for tests and evaluation. However, with the cessation of hostilities large scale deliveries of this type never materialised.

428. Standard production M26 Pershing.

429. T92 HMC, first pilot model.

WITH acceptance of the T26E3 heavy tank in January 1945, it was proposed to develop a complete "Heavy Weight Combat Team", a series of complementary AFV types based on a common chassis and its component parts, that of the T26E3. The towed 240mm howitzer M1918 had not proved entirely satisfactory, due mainly to the difficulty of hauling a gun of this great weight across country, and experience with the 155mm gun mounted on the M3/M4 series chassis indicated that it would be feasible to mount the 240mm howitzer on a chassis based on that of the T26E3 heavy tank. This carriage would also serve to mount the 8in gun. With the 240mm howitzer the project was designated T92 HMC, while the same chassis with the 8in gun was designated T93 GMC. Other members of the "Heavy Weight Combat Team" were to be the T26E3 heavy tank, the T26E5 assault tank, the T84 8in HMC, the T31 cargo carrier, and (later) the T26E2 (M45), all described in the previous section.

Design of the T92 was approved in March 1945 and the project was classified "limited procurement", a contract for four pilot vehicles being placed immediately with Chrysler (Detroit Arsenal). First of these vehicles was completed early in July 1945, and after initial trials at APG was delivered to Fort Bragg for Artillery Board tests. The T93 was ordered at the same time and classified as a "limited procurement" type in April 1945. Two of the four pilot models of the T93 had been completed and delivered by September 1945, when all orders were cancelled on cessation of hostilities with Japan. Total output of T92s amounted to five (including pilots) with just the two T93s, all built by Detroit Arsenal. Trials indicated that these heavy weapon carriages would be ideal for bombardment work in the bunker and cave warfare type of fighting being experienced against the Japanese in the Pacific. At the time of the Japanese surrender the T92s and T93s were being prepared for shipment to the Pacific area for use in the planned invasion of Japan. Shells with concrete-piercing fuses were developed specially for the T92 and T93.

To mount these large calibre weapons it was necessary to lengthen the T26E3 chassis and add an extra bogie wheel each side. Chassis layout was also reversed so that drive was now taken to a front sprocket. A recoil spade was fitted at the rear.

430. T93 GMC, first pilot model.

HMC, T92

SPECIFICATION:

Designation: 240mm Howitzer Motor Carriage T92
8in Gun Motor Carriage T93
Crew: 8 (driver, co-driver, commander, gun crew (5))
Battle weight: 137,500lb (T92), 131,400lb (T93)
Dimensions: Length 28ft (excluding gun) Track width 23in
Track centres/tread 9ft 2in.
Height 10ft 8in (T92)
8ft 5in (T93)
Width 11ft
Armament: 1 × 240mm howitzer M1 (T92)
1 × 8in gun M1 (T93)
Armour thickness: Maximum 25mm
Minimum 13mm
Traverse: 12° right to 12° left. Elevation limits: +65° to 0°
Engine: Ford GAF V-8 petrol
470hp
Maximum speed: 15mph
Maximum cross-country speed: —
Suspension type: Torsion bar
Road radius: 50-80 miles
Fording depth: 4ft 7in
Vertical obstacle: 3ft 10in
Trench crossing: 7ft
Ammunition stowage: 6 rounds in vehicle; additional rounds in accompanying cargo carrier (eg, M30 or T31)
Special features/Remarks: T93 is distinguished from T92 by longer barrel of 8in gun which also accounts for weight difference. Otherwise both vehicles are identical and mechanically similar to the M26 series.

431. T93 GMC emplaced for firing with recoil spade lowered.

432. Top detail view of T92 HMC with howitzer in travelling position.

433. Pilot model of the T95 GMC.

IN September 1943 the Ordnance Department suggested the construction of a very heavily armoured tank, the T28, for attacking heavily fortified enemy positions and invulnerable against all known enemy tanks. It was to have an armour basis of 8in and be based on the chassis and mechanical components of the T23 medium tank. It was proposed to mount the newly developed 105mm gun T5E1, which was a high velocity weapon. Army Service Forces gave approval for development in April 1944 but stipulated that mechanical drive was to be used rather than the electric transmission of the T23. Five pilot models of the finalised design were ordered from Pacific Car & Foundry, but in March 1945, while work was under way on the first vehicle, the designation was changed to 105mm Gun Motor Carriage T95 since the gun was not mounted in the turret but in the hull. With the cessation of hostilities against Japan the order for pilot models was reduced to two only, the first being completed in September 1945. Trials with the pilot models thus took place after the war, the second pilot being subsequently destroyed by fire during one of its trial runs. All work on the project terminated in October 1947 due to the successful development of the T29, and no production orders were placed, though quantity production had at one stage been contemplated should the war against Japan have continued.

The T28/T95 was the heaviest American AFV design of the war and also one of the most unusual. In some ways it resembled the British Tortoise (qv), but exhibited some more novel features. The hull was a cast structure two-thirds the length of the track assemblies and set towards the rear. It had a jib at the rear for loading the ammunition into the fighting compartment. The turtle shaped superstructure featured a vision cupola for the commander surmounted by an AA machine gun on a ring mount. The 105mm gun was set in the hull front with limited traverse in a ball shaped mantlet of 12in armour. Each track assembly was a twin unit made up of two complete HVSS units, the outer of which on each side could be disconnected and removed to reduce the vehicle's width and weight for transportation by rail (or for road running in confined surroundings). The two detached track units could then be linked together side by side to form a "dumb" unit which could either be towed by the vehicle itself or by an attendant prime mover. Small jibs were provided on the vehicle to assist in detaching the outer tracks.

SPECIFICATION:
Designation: Heavy Tank T28 (later: 105mm Gun Motor Carriage T95)
Crew: 8 (commander, driver, co-driver, gun crew (5))
Battle weight: 190,000lb
Dimensions: Length 36ft 6in (overall) 24ft 7in (excluding gun)
Track width 19½in (each track unit)
Track centres/tread 10ft 1in
Height 9ft 4in
Width 14ft 5in
13ft 11in (outer tracks removed)
Armament: Main: 1 × 105mm T5E1 gun
Secondary: 1 × ·50 Browning MG (AA)
Armour thickness: Maximum 300mm
Minimum 25mm
Traverse: 10° right and 10° left. Elevation limits: +19½° to −50°
Engine: Ford GAF V-8 petrol. 410hp
Maximum speed: 8mph
Maximum cross-country speed: —
Suspension type: HVSS on four separate track units (64 bogie wheels)
Road radius: 100 miles
Fording depth: 3ft 11in
Vertical obstacle: 3ft
Trench crossing: —
Ammunition stowage: 62 rounds 105mm
660 rounds ·50 cal
Special features/remarks: Massive heavily armoured vehicle of limited tactical value due to its great weight and slow speed. Outer track units detachable. Rear drive with Torquematic transmission, similar installation to M26 heavy tank.

434. Heavy Tank T32.

Heavy Tank T32 and T32E1: This was an improved version of the T26E3 (M26) designed to provide better armour protection without impairing the performance or reliability of the M26. Hull was the same as that of the M26, lengthened by one bogie wheel each side and with armour maximum increased to 125mm at the front and 75mm at the sides. Turret had 200mm frontal armour. Improved T15E1 90mm gun was fitted and a counterweight was added on the turret rear. Engine was uprated to 750HP and cross-drive transmission replaced the Torquematic transmission of the M26. Apart from increased length and increased weight, other details were as for M26 series vehicles. A batch of pilot models was ordered from Chrysler (Detroit Arsenal) in February 1945, but were not completed until early 1946. No production order followed. The T32E1 was similar to the T32 except that it had a welded hull front instead of the cast front, while the hull machine gun was eliminated.

435. Heavy Tank T29.

Heavy Tank T29: Development of this vehicle started in March 1944 in an attempt to produce a heavy tank with firepower and armour protection superior to that of the T26E3 (M26). It was intended to fit the Cross-Drive transmission and a Ford tank engine uprated to 750HP. Approval for building pilot models was given in September 1944. Hull was similar to that of the T26E3 but lengthened to take a massive new cast turret to hold the 105mm T5 gun. The General Staff authorised production of this type in February 1945 for use in the war against Japan where heavy calibre weapons were considered necessary for firing against bunkers and caves. Army Ground Forces, however, were opposed to vehicles as large as this and stated that they had no requirement for them. With the cessation of hostilities, production was limited to a batch of pilot models only for testing and development. These were delivered in 1947. Details: combat weight: 138,000lb; crew: 6; armament: 1 × 105mm gun T5; length: 25ft (excluding gun); width: 12½ft; height: 10ft 7in; top speed 18½mph; ammunition stowage: 63 rounds; engine: Ford 750HP.

436. Heavy Tank T30.

Heavy Tank T30: This was a parallel design to the T29 evolved and produced at the same time and within the same programme. Principal difference was the installation of a Continental 810HP air-cooled engine in place of the Ford unit, and the mounting of a 155mm gun T7 in place of the 105mm weapon. This vehicle included a rammer in the turret for loading the gun which fired separate ammunition. Both the T29 and T30 were classified "limited procurement" types in April 1945. T30 details as for T29, except: armament: 1 × 155mm gun T7; combat weight: 144,500lb; ammunition stowage: 34 rounds; top speed: 16½mph; engine: Continental 810HP.

437. Heavy Tank T34.

Heavy Tank T34: This resulted from the adaptation of the standard American 120mm AA gun to a form suitable for mounting in a tank. The design of the T29/T30 series was modified to take the 120mm gun T53 by suitable changes in the gun mount, but with no fundamental alterations to the basic design. One T29 and one T30 pilot model were each fitted with the 120mm gun and re-designated as Heavy Tank T34. Approval for this development was given in April 1945, but the pilot model T34s were not delivered until 1947. No production orders followed but the post-war M103 heavy tank design stemmed from the T34. Details as for T29/T30 except for 120mm gun.

438. 3in GMC T1(M5).

3in Gun Motor Carriage T1 (M5): This vehicle was based on the chassis of the commercial high speed tractor produced by Cleveland Tractor (Cletrac). Development started in December 1940, the specification calling for a 3in gun in an open shield mounted at the rear with limited traverse. The crew manned the gun from the ground. Modifications to the design in March 1941 called for a bigger shield. The pilot model was completed in November 1941 and standardised as the T5 GMC. However, no production orders were given, since far superior SP designs had been produced in the meantime. In September 1942 the T1 (M5) was declared obsolete. The T1 GMC had a 160HP Hercules diesel engine.

439. Bigley Gun Motor Carriage.

Gun Motor Carriage T42 (Bigley Gun Motor Carriage): This was a commercial development of the Christie M1937 chassis fitted with a new turtle backed superstructure of duralumin with a mock-up 37mm gun and MG in the glacis plate. It was offered to the US Army by Bigley as a prototype fast tank destroyer. Various other armament installations were proposed, including a rear firing MG and two side firing MGs. It was tested briefly in 1941-42 under the designation T42 GMC, but was rejected as unsatisfactory for army requirements.

MISCELLANEOUS PRIME MOVERS AND CARGO CARRIERS BASED ON TANK COMPONENTS

United States

440. Cargo Carrier T22.

Cargo Carrier T22: This was the proposed ammunition carrier to accompany the 4·5in Gun Motor Carriage T16 (see plate 246). It utilised the same chassis as the T16 with a cargo compartment replacing the gun mount. Development started in March 1943. When it was proposed to utilise the T24 (M24) light tank chassis for the T16, the modified design being designated T16E1, the design of the T22 was similarly changed to use this chassis under the designation T22E1. In January 1944, when the T16 and T16E1 projects were dropped, the same fate awaited the T22 and T22E1. Only pilot models of the T22 were built.

441. Heavy Tractor T2.

Heavy Tractor T2: Development of this vehicle started early in 1941 utilising the suspension and other chassis components of the M2A1 medium tank with various minor alterations. Power was provided by a GM diesel engine and the vehicle was to be used for hauling the 155mm gun and 8in howitzer. In June 1941 an improved design, the T12, was authorised as a development of the T2. The T2 proved unsatisfactory, however, and both the T2 and T12 were abandoned in June 1942 in favour of later designs. The T2 pilot model was subsequently used for transmission experiments.

13 ton High Speed Tractor M5 (formerly Medium Tractor) (T21): International Harvester began development of a medium weight tractor suitable for hauling field artillery of 105mm and 155mm types at the end of 1941. Two alternatives were built, identical except for tracks and suspension. The T20 had rubber band tracks and was ultimately rejected in favour of the alternative version, T21, which had tracks and suspension from the M3 light tank. It was standardised in October 1942 as Medium Tractor M5. Produced in quantity it was one of the most important US artillery tractors. Details: weight: 28,300lb; crew: 9; length: 15ft 11in; height: 8ft 8in (over canopy); width: 8ft 4in; maximum towing speed: 35mph; engine: Continental R-6572, 235HP. The vehicle had a front-mounted winch and air braking equipment for the tow. Later models had a ·50 cal MG in a ring mount behind the canopy. The M5 was re-classified as a High Speed Tractor in August 1943.

442. 13 ton High Speed Tractor M5.

18 ton High Speed Tractor M4 (formerly Medium Tractor) (T9): Development of Medium Tractor T9 commenced in June 1941 to provide a vehicle able to tow artillery equipment up to 30,000lb in weight. It was standardised and ordered into production in August 1942, being built by Allis-Chalmers. It was a comfortably equipped vehicle with full weather protection and separate driving and crew compartments. The chassis utilised components from the M2A1 medium tank. Built in quantity, it was used to tow the 90mm AA gun, 155mm gun, and 8in and 240mm howitzers. Vehicles built as towers for the 90mm AA gun (Class A) had different stowage arrangements to vehicles used with other equipment (Class B). Details: weight: 31,500lb; crew: 11; length: 16ft 11 in or 17ft 2in (Class A); height: 7ft 10in; width: 8ft 1in; maximum towing speed: 33mph; engine: Waukesha 145GZ, 210HP. The vehicle had air and electric braking equipment for the tow, a rear-mounted winch, and a folding ammunition hoist in Class B configuration. A ·50 cal Browning MG fitted in a ring mount on the roof.

443. 18 ton High Speed Tractor M4 towing 8in gun.

Note: Other prime movers, tractors, and cargo carriers which were directly based on tank chassis are described with other variants under the heading of the basic vehicle. Towing vehicles not based on tank components are omitted.

444. 18 ton High Speed Tractor M4, front view.

Other Special Purpose Vehicle Projects

Several projected special purpose vehicles or equipments were under development at the cessation of hostilities. Prototypes did not appear until after the war's end. Most important of these included:

Tank Recovery Vehicle T12: Recovery vehicle based on the M26 heavy tank with modified turret, winch, and boom. Design began in late 1944.

Tank Recovery Vehicle T6E1: Recovery vehicle based on the chassis of the M24 light tank. Design started in September 1943, but the T6E1 was held for possible production only if the need arose.

Mine Resistant Vehicle T15E3: As T15/E1/E2, but based on T26E3 chassis instead of M4 medium tank. 1945 design.

155mm Mortar Motor Carriage T90 and T96: Breech loading mortar on chassis of M4 medium tank and M37 HMC respectively. Design started December 1944. Cancelled, August 1945.

445. Pilot model T6E1 was a proposed recovery version of the M24 light tank.

446. LVT(A) 4 showing 75mm howitzer.

THE American LVTs stemmed from the Alligator design of Donald Roebling Jr, who had produced a lightweight tracked amphibious vehicle in 1935 for rescue work on the swamps of Florida Everglades. This attracted the attention of the USMC and in 1940 Roebling re-designed the vehicle to suit Marine requirements. First order, for 200 vehicles, designated LVT 1 (landing vehicle tracked) was placed in November 1940 and in August 1941 the first USMC amphibian tractor battalions were formed with these vehicles. Larger improved version was designated LVT 2. Both these types were essentially cargo carriers for ship-to-shore supply operations and were made of mild steel. As such they are beyond the scope of this book.

In November 1943, LVT 2s were used to carry troops ashore at Tarawa and it was realised that the LVT suitably modified could be used to provide fire support for amphibious operations. What was virtually a floating tank was therefore developed, simply by building an LVT 2 in armour plate instead of mild steel, decking in the cargo space and adding the turret and 37mm gun from the M3 light tank. This became the LVT(A) 1, (A: armoured). Developments from then on included the following:

LVT(A) 1–Landing Vehicle Tracked (Armored) Mk I: This was the LVT 2 design modified as described to form a hybrid amphibious tank. Designed at the end of 1943, the LVT(A) 1 was in service in 1944 and proved most successful. In addition to the turret, two machine gun positions were cut in the rear decking with ·30 cal Brownings mounted on Scarf rings and fitted with shield and coamings. Weight: 32,800lb; length: 26ft 1in; height: 10ft 1in; width: 10ft 8in; crew: 6; maximum speed (land): 25mph; maximum speed (water): 6½mph; armament: 1 × ·37mm gun M3, 1 × ·50 cal MG (AA), 2 × ·30 cal MG; ammunition: 104 rounds 37mm, 6,000 rounds ·30 cal; engine: Continental W-670, 250hp; track width: 14¼in.

LVT(A) 2: Built concurrently with the LVT(A) 1, this vehicle was identical to the original mild steel LVT 2 in appearance but was constructed of armour plate like the LVT(A) 1 with the turret, guns, and decking omitted so that it could be used as an armoured troop/store carrier for assault landings to carry troops over the beach and inland for disembarkation. The idea for this came after the Tarawa landings at the end of 1943, when the original LVT 2 had been given extemporised armour protection from bolted on plating. The LVT(A) 2 was built to US Army requirements. Details as for LVT(A) 1 except for omission of armament and turret, etc. These basic LVTs were called "Water Buffalo" by the American forces.

LVT(A) 4: In March 1944, by which time the need for heavier armament than the 37mm gun had been appreciated, the LVT(A) 1 was further modified by the substitution of the complete turret and 75mm howitzer from the M8 GMC (qv). The machine gun positions were eliminated due to the extra weight. Details as LVT(A) 1, except: weight: 40,000lb; height: 10ft 5in; armament: 1 × 75mm howitzer M2 or M3, 1 × ·50 cal MG (AA); ammunition: 100 rounds 75mm, 400 rounds ·50 cal.

LVT(A) 5: The LVT(A) 4 proved a most useful and successful vehicle in the many amphibious assault landings of the Pacific war. It suffered however, by having only hand traverse and no stabilisation of the gun. LVT(A) 5 remedied these shortcomings by the addition of power traverse and a gyro-stabiliser in the turret. Designed in 1945, it did not come into service, however, until after the war. Details and appearance as LVT(A) 4.

LVT 4: A disadvantage of the original LVT 1 and 2 designs had been the rear mounted engine and central cargo space which meant that troops and stores were loaded over the side of the vehicle and that there were thus limitations on the sort of item which could be loaded. The LVT 4 was basically the LVT(A) 2 modified by having its engine moved forward and resited immediately behind the driving compartment. The transom was then replaced by a ramp operated by a hand winch. This now allowed troops and stores to be loaded through the stern of the vehicle. It could carry 30 troops (compared to 18 in the LVT 2) and light vehicles (eg, Jeep, Universal Carrier) or field guns. This type was first used at Saipan in mid 1944, and was also used in Italy and NW Europe, 1944-45. The LVT 4 was also used by the British Army, under the designation "LVT, Buffalo". In British service it was fitted with a Polsten 20mm cannon and two ·30 cal Browning MGs. In American service it carried a mount on each side for either a ·30 cal or ·50 cal machine gun. These could be seen fitted with or without gun shields. The British army, incidentally, also received a very small batch of LVT 1s (Alligators) which were used mainly for crew training. LVT 4 details as LVT(A) 4 except: weight: 33,350lb; height: 8ft 1in; armament as stated above.

LVT 3: This was vehicle developed by Borg Warner which was similar in external appearance to the LVT 4, complete with stern ramp, but which had a single Cadillac 125HP engine mounted in each side pontoon and the Hydra-matic automatic transmission used in the M5 light tank (qv). This produced a vehicle of superior performance, more efficient than the Continental-engined designs. Called the "Bushmaster", it was produced in 1944 and first used in action at Okinawa. Details as LVT 4 with differences as noted above.

LVT(A) 4 with M24 turret: This was an experimental conversion, produced in January 1945 to provide an even more heavily armed LVT for support operations. The M8 type turret was replaced with that from a Chaffee M24 light tank and suitable alterations were made to the super-structure to allow the larger turret to be fitted. No further development took place after the cessation of hostilities.

Rocket launchers and flame-projectors were fitted to some LVT models and tested in combat. Two examples are shown.

Rocket launcher T45 mounted on stern of LVT(A) 4: Could also be mounted on stern of LVT(A) 2 and LVT 1. Two banks of 10 rockets.

Other rocket launcher types on LVTs: **LVT 4 with T54 rocket launcher**—this consisted of 20 7·2in rockets in square sectioned tubes in a heavy framed carrier, mounted centrally in the cargo space to fire forward.

LVT 3 with T89 rocket launcher: This was a similar arrange-ment to the LVT 4 with T54 RL. It had 10 7·2 in rocket tubes.

E7 flame-gun in LVT(A) 1: The flame-gun replaced the 37mm gun and the fuel tank was carried internally. Same equipment could be fitted in the LVT(A) 4 in place of the 75mm howitzer, and also in the LVT 4, with the projector mounted in a small shield in the centre front of the cargo space. All developed and fitted in 1944.

Other flame-thrower equipment fitted to LVTs included the Canadian Ronson which could be mounted in the LVT(A) 4 in place of the 75mm howitzer, or in the LVT 1 and 2 in the left hand front of the superstructure front.

LVT(A) 1 with E14-7R2 flame-thrower: This was an early adaptation with the projector replacing the 37mm gun. Ten were built in 1942 but were not used in combat.

LVT(A) 2 with portable ramp: This was a specialised local modification to a few vehicles whereby an Ark type bridge was fitted to the top of a standard LVT to enable troops and equipment to surmount coral cliffs during the Tinian landings.

Note: All LVT types were propelled in the water by cup-shaped grousers which formed an integral part of the track shoe. Suspension was achieved by rubber cushioning of the bogie wheel axles. Original US Army designation for the LVT 1 was Amphibian Tractor T33, but later this was dropped and the US Navy LVT

447. LVT(1) showing 37mm gun.

designations were adopted. LVTs were used by both the US Army and USMC. They were a US Navy responsibility. Turreted versions carried bouyancy bags to compensate for extra weight. Details applicable to all variants: Range (land): 125 miles; Range (water): 75 miles; Armour (maximum): 13mm; (minimum): 6mm; Vertical obstacle: 3ft; Trench crossing: 5ft; Minimum freeboard: 2ft 3in.

450. LVT(A) 4 prototype fitted with M24 turret.

448. LVT 4 Buffalo with machine guns in shields.

451. T45 Rocket Launchers mounted on a LVT(A) 4.

449. Borg Warner "A" vehicle 1942. Though this prototype vehicle did not go into production in this form, it led to the LVT 3 Bushmaster which was of similar appearance but with rear ramp and without the turret.

452. E-7 Flame-thrower on LVT(A) 1.

453. LVT(A) 1 firing at night on practice ranges.

PART 3

COMMONWEALTH VEHICLES

454. Although Australia, Canada, and New Zealand all built indigenous tank designs, only the Canadian tanks reached production status. Meanwhile, all Commonwealth armies used British and American tanks. These are Light Tanks Mk VIA of the Australian Light Horse (3rd Australian Tank Corps) in training at Rokeby, the Australian Tank Corps depot.

455. Ram Mk I, 2pdr gun and side doors.

PRIOR to the outbreak of war it was considered that Canadian industry did not lend itself to tank production and it was not anticipated that war requirements would include the building of tanks in that country. In the event of war, Canadian tank units would have been issued with British tanks.

German military success in the invasion of Poland—and subsequently France and Flanders—however, showed that British tank production would have to be greatly expanded. In the spring of 1940, therefore a British order for Valentine tanks (qv) was placed with Canadian Pacific, and in the light of this a Canadian order was immediately placed with the same contractor to equip a Canadian Tank Brigade. The fall of France, the danger of invasion in Britain, and the German bombing offensive, led to a further decision that more tank and military production should be established in Canada and the formation of two Canadian armoured divisions on the summer of 1940 added a further requirement for 1,000 cruiser type tanks to equip them. The serious shortage of tanks in Britain meant that there was no chance of any such equipment being provided from the United Kingdom. At the same time, American tank production, though about to be increased was insufficient to provide Canada's needs.

Hence a decision was taken to construct a Tank Arsenal in Canada under the administration of Montreal Locomotive Works with the assistance of its parent organisation, American Locomotive. It was further decided that the Canadian-built cruiser tank design would be based on that of the US M3 medium tank (qv) then at the pilot model stage, both to save time and to utilise mechanical and chassis components already in production for the M3. By the autumn of 1940, it became clear that many of the design features being finalised and approved for the M3 medium by the US Ordnance Department would be far from satisfactory for British and Canadian users, in particular the high silhouette, sponson-mounted main armament, inadequate armour protection, and lack of radio in the turret.

After lengthy consideration, therefore, the Inter-departmental Tank Committee decided, in January 1941, to continue the Canadian design utilising the mechanical components of the M3 medium but incorporating hull and turret to suit Canadian requirements, with a British main armament. This vehicle was to be known as the Ram, the name being derived from the family crest of General Worthington, the chief of Canadian armoured forces. Responsibility for development was placed with the Montreal Locomotive Works directed by the Canadian Department of Munitions and Supply, with the British Tank Mission in America and the Department of National Defense acting in an advisory capacity.

A running prototype was completed in June 1941 and design and stowage details were worked out during the course of the year. Procurement of suitable armour plate in Canada and its heat treatment, forming, and machining, presented an immense problem which was eased by the provision of the turret and upper hull castings from General Steel Castings in USA. It was planned to provide 6pdr gun mount drawings from Britain but these did not materialise in time and the mantlet, cradle, and elevating gear were designed in Canada. Pending the final development and production of the 6pdr mountings, the first fifty vehicles were fitted with 2pdrs and were designated Ram Mk I Production vehicles with 6pdrs then became Ram Mk II, the latter achieving full production status in January 1942. In the summer of 1941 the Ram pilot model was loaned to the US Army, carrying out running trials at APG with a British crew under an officer from the Ordnance Department. The T6 (M4 medium tank prototype) had several features similar to the Ram, notably the hull shape and the sponson doors. Much of the similarity was co-incidental, however, the T6, like the Ram, being developed to overcome the inadequacies of the M3 medium tank.

Rams were not used in action as gun tanks, most being used for training only, in Canada or Britain. Many were subsequently converted for special purposes—notably as Kangaroo APCs—as noted below. Ram Kangaroos equipped the armoured troop carrier battalions of 79th Armoured Division in the NW Europe campaign and were the first tracked APCs used in quantity by the British. These vehicles remained in service for some years post-war until specialised APCs were developed.

The Ram chassis resembled and followed the design of the American medium tank with vertical volute suspension. Lower hull was of riveted armour plate and upper hull and turret were of heavy cast armoured steel. Engine was a Continental R-975 as for the M3 medium. First 50 vehicles were Mk I with 2pdr guns. Remainder were Mk II with 6pdrs. Ram II had gyro-stabiliser for main armament and splash beading round turret ring. Turret was hydraulically traversed. Track was of a pattern designed in Canada, known as CDP type, and was generally considered superior to the steel and rubber types produced for the M4 mediums, being easier and cheaper to produce and offering better traction.

Several improvements and modifications were incorporated in the Ram during production. These included the addition of a transmission oil cooler, elimination of the sponson doors, elimination of the machine gun cupola, addition of a floor escape hatch, elimination of pistol ports in turret, and modifications to the engine to allow use of 80 octane fuel. Total Ram II production was 1,094 vehicles.

SPECIFICATION:

Designation: Tank Cruiser, Ram Mk II or Mk I (US designation, M4A5).
Crew: 5 (driver, co-driver, gunner, loader, commander)
Battle weight: 65,000lb
Dimensions: Length 19ft; Height: 8ft 9in; Width 9ft 1in; Track width $15\frac{1}{2}$ in (with CDP tracks); Track centres/tread 6ft 11in
Armament: Main: 1 × 2pdr OQF (Ram I); 1 × 6pdr OQF (Ram II)
Secondary: 2 × ·30 cal Browning MG; 1 × ·30 cal Browning MG (AA)
Armour thickness: Maximum 87mm; Minimum 25mm
Traverse: 360°. Elevation limits: $+20°$ to $-7\frac{1}{2}°$
Engine: Continental R-975. 400hp.
Maximum speed: 25mph
Maximum cross-country speed: 20mph
Suspension type: Vertical volute
Road radius: 144 miles
Fording depth: 3ft
Vertical obstacle: 1ft 6in
Trench crossing: 7ft 5in
Ammunition stowage: 92 rounds 6pdr (Ram II), 171 rounds 2pdr (Ram I); 4,440 rounds ·30 cal
Special features/remarks: Vehicle could also be fitted with American pattern $16\frac{1}{2}$in wide tracks as used on M4 medium. OP version had dummy gun with limited turret traverse, by hand, 45° each side. Rams could be seen fitted with prominent stowage box on turret rear. Excellent design, in many ways superior to the early M4 which over-shadowed it by being produced in much greater quantity.

456. Ram Mk II, 6pdr gun.

457. Ram Mk II, late production vehicle, sponson doors eliminated.

458. Ram OP/Command; note cable reels on rear decking.

459. Ram ARV Mk I recovering Ram Mk II. Note comparison between Mk I and Mk II turret fronts, and tool boxes on rear of ARV.

RAM

VARIANTS

Ram Kangaroo: Turret and associated equipment removed to provide space for carriage of infantry, 11 men in battle order plus crew of two. Hand and foot grips added to each side of the hull.

Ram Gun Tower: As Kangaroo but with towing hook at rear for 17pdr anti-tank gun. Crew and ammunition carried in vehicle.

Ram Ammunition Carrier: Also known as the Wallaby, this was converted as for the Kangaroo, but was fitted with racks for 25pdr ammunition as limber vehicle for the Sexton (qv).

Ram OP/Command: Ram II designed for use as armoured observation post or command vehicle. Fitted with dummy gun and extra radio equipment, aerials, and telephone reels, etc. 84 built in 1943.

Ram GPO: As Ram OP, but fitted with special equipment for Gun Position Officers of SP artillery regiments. Included Tannoy loudspeakers.

Ram ARV Mk I: Ram Mk I converted to ARV by addition of winching gear.

Ram ARV Mk II: Ram II with turret replaced by fixed dummy turret and dummy gun, with demountable forward jib, fixed rear jib, and earth spade. Appearance and specification as for British Sherman ARV Mk II (qv).

Ram AVRE: Two vehicles converted by Royal Engineers in 1943 to examine possibility of using Rams in the assault engineer role. The Churchill was adopted as standard AVRE and the Ram was abandoned for the purpose.

Ram QF 3·7in AA: Attempt in late 1942 to produce a self-propelled mount for the 3·7in AA gun. Various modifications (including shield) made during its trials. Project abandoned after test.

Ram Flame-throwers: A small number of Kangaroos converted by Canadian Army to include Wasp II flame-throwing equipment. Also known as the Badger, though most vehicles so converted were Shermans.

Ram with 75mm gun: Trials were carried out on one vehicle with American 75mm M3 gun replacing the 6 pounder. Project abandoned.

Ram Searchlight: Local conversion in 1945 whereby a Ram Kangaroo was fitted with a complete 40in searchlight and generator to illuminate airstrips, etc, in forward areas for night operations, NW Europe.

460. Ram QF 3·7in SP AA gun, seen on trials.

461. Ram Kangaroo APC of 79th Armd Div, April 1945. Note US tracks (M4 type) on this vehicle.

462. Rear view of Ram Mk 1.

463. Skink AA Tank.

THE Grizzly I was the basic M4A1 medium tank design built by Montreal Locomotive Works after completion of the Ram II and Ram OP/Command contracts. It differed from the M4A1 principally by being fitted with Canadian CDP tracks (though M4 tracks could also be used) and by the installation of British pattern wireless equipment as built. A prominent stowage box was added on the turret, and sandshields were fitted as standard. Another feature introduced in the Grizzly was a 2in smoke mortar in the turret roof to meet British/Canadian requirements. This was, as a result of experience with the Grizzly, subsequently fitted in all American-built M4 series tanks. Plans to produce the M4A1 in Canada had been made as early as September 1942. However, the limited facilities of Montreal Locomotive Works meant that production could not start for another year. Between September-December 1943, a total of 188 Canadian-built M4A1s was produced, the name Grizzly I being given to this vehicle by the Canadians. M4A1 production in Canada was prematurely cut short by the decision to concentrate all M4 series production in American plants, which in 1944 had sufficient free production capacity to meet all M4 series tank requirements for Allied armies. The Grizzly, together with US-built M4s, equipped Canadian armoured battalions in Canada and Europe, 1944-45.

Details as American M4 series (qv) except: combat weight: 67,000lb; tracks: 15$\frac{1}{4}$in (CDP) or 16$\frac{1}{2}$in (American type); length: 19ft 1in (excluding sandshields); height: 9ft 10in; ammunition stowage: 78 rounds 75mm.

VARIANT

Skink AA Tank: In the autumn of 1943 when there was a requirement for AA tanks for the forthcoming invasion of Europe, the Department of National Defense contracted with the Waterloo Manufacturing Co to develop such a vehicle on the basis of the Grizzly tank. This firm designed and built a modified turret, dimensionally similar to that of the Grizzly, which mounted four 20mm Polsten cannon and associated sighting equipment. Production orders were given in early 1944 after approval of the pilot model, but in mid 1944, due to proven Allied air superiority in Europe, the requirement for AA tanks lapsed, and orders were cancelled after only a few Skinks had been completed. One of these was tested in Britain. Details as for Grizzly except for 4 × 20mm Polsten cannon which fired singly, in salvoes of two, or all together. Magazines for re-loads were carried in armoured bins which replaced the ammunition racks of the original Grizzly.

464. Grizzly cruiser tank—Canadian-built version of M4A1 medium tank.

SELF-PROPELLED GUN, SEXTON **Canada**

465. Early production Sexton with three-piece nose and M3 type bogies.

ESSENTIALLY an "anglicised" version of the American M7 (Priest), the Sexton was developed in the latter half of 1942 to meet British General Staff requirements for a self-propelled gun with all the mobility and characteristics of the M7 HMC (qv) but with the standard British 25pdr field howitzer in place of the American 105mm weapon. As a basis, the Ram chassis (itself fundamentally the US M3 medium tank chassis) was used. Layout was similar to that of the M7, but the driving position was on the right and the gun was offset to the left, since the Ram chassis had the driver's position shifted to the right hand side. A small ammunitioning hatch was provided in the left hand side, and there was no ring mount for an AA gun as in the M7. The standard 25pdr howitzer was mounted, but in order to give sufficient elevation it was necessary to limit the recoil throw. Built by Montreal Locomotive, the pilot model was shipped to Britain for tests and approval, and production commenced at Montreal Locomotive Works in early 1943, 424 vehicles being completed by the year's end. Successive orders from Britain, kept the vehicle, named Sexton, in production until the end of 1945, by which time 2,150 had been completed.

The Sexton chassis was identical to the Ram in early production vehicles, with a welded superstructure. Changes introduced during production included the adoption of a one-piece cast nose (as on late M4 medium tanks), M4 type bogies with trailing return rollers, a towing hook at rear for ammunition trailer, provision of an auxiliary generator, stowage for extra equipment, and mounts for Bren AA machine guns.

The M7 HMC had entered service with the British in October 1942 under Lend-Lease arrangements. As the Sexton became available it gradually replaced M7 Priests in the field regiments of armoured divisions, the M7 being entirely superseded by mid 1944 (see M7 history). Sextons remained in British and Canadian service for many years post-war.

SPECIFICATION:

Designation: 25pdr SP, tracked, Sexton
Crew: 6 (commander, driver, gunner, gun-layer, loader, wireless operator).
Battle weight: 57,000lb
Dimensions: Length 20ft 1in; Height 8ft; Width 9ft; Track width 15½in. (or 16½in with American tracks); Track centres/tread 6ft 11in
Armament: Main: 1 × Ordnance 25pdr howitzer Mk II
Secondary: 2 × Bren ·303 cal MG (AA)
Armour thickness: Maximum 25mm
Minimum 12mm
Traverse: 25° left to 15° right. Elevation limits: +40° to −9°
Engine: Continental R-975 radial air-cooled 400hp
Maximum speed: 25mph
Maximum cross-country speed: 20mph
Suspension type: Vertical volute
Road radius: 125-145 miles
Fording depth: 3ft 6in
Vertical obstacle: 2ft
Trench crossing: 7ft 5in
Ammunition stowage: 87 rounds 25pdr (HE or Smoke)
18 rounds 25pdr (AP)
50 Bren gun magazines
Special features/remarks: Successful expedient design, though under gunned for the size and power of the chassis. Layout based on that of M7 HMC. Built to British specifications, featuring British wireless (No 19) set, and Tannoy. Howitzer mounted in prominent armoured cradle.

VARIANT

Sexton GPO: Introduced in late 1943 (the prototype was designated "G" vehicle), this was a Sexton with gun removed and extra equipment added for gun position officers (GPO) of Sexton batteries. It was fitted with extra radio, map tables, extra telephone cables, and an extra Tannoy unit.

466. Late production Sexton with one-piece cast nose and M4 bogies with trailing return rollers.

OTHER TANKS AND SP TYPES USED BY THE CANADIAN ARMY, 1939-45

Light Tanks Mk VI, VIA, VIB: Small number supplied for training in 1940, used at the Canadian Armoured Corps depot, Camp Borden. For details of these vehicles see British light tank section.

Valentine VI, VII, VIIA: Canadian-built models of the Valentine, produced by Canadian Pacific. 1,420 were built, but all except 30 were shipped under Lend-Lease to Russia. Remainder used in Canada for training. Originally planned as major type for Canadian use, but supplanted by the Ram as a service type in Canada. For Valentine details see British infantry tank section.

M4 Medium Tank (Sherman): All models, but mainly M4A1, M4A2, M4A4. For details of M4 series see American medium tank section.

M10 GMC (Wolverine, Achilles): For details of M10 series see American medium tank section.

Sherman Kangaroo, Priest Kangaroo: For details of these variants see M4 series, British Service, and M7 HMC, British Service.

Sherman Badger: Flame-thrower conversion of M4A3 (HVSS), 1945, with turret removed. See M4 series, British Service.

Churchill I, III, IV, V, VII: For details of these vehicles, see British infantry tank section. Widely used by Canadian forces in Europe, 1942-45. First used in action (at Dieppe) by Canadian regiments.

Tank Mk VIII: In order to provide Canadian forces with tanks for training purposes in summer 1940 at a time of acute shortage, the United States sold 229 1919 vintage tanks to the Canadian government at nominal scrap value. Of these, about 90 were Mk VIII tanks which had been in store since the early thirties after withdrawal from US Army service. These were reactivated and delivered in September 1940. Totally obsolete, they were used only for training at Camp Borden until sufficient modern tanks became available.

6 Ton Tank: Balance of the 229 obsolete tanks sold to Canada in 1940 were the American-built Renault FT type which were two-man machines of limited performance. Used for training at Camp Borden, they were discarded when modern types became available.

467. Five valentines Mk VI after completion at Montreal Locomotive Works pose beside a Canadian Pacific locomotive built in the same plant. Note the cast nose, the distinguishing feature of all Canadian-built Valentines.

468. Light Tanks Mk VIB in use for training at Camp Borden, the Canadian armoured corps centre, early in 1941.

469. Mk VIII tanks, shown on delivery to Canada, September 1940.

470. Canadian crews in training with ex-American 6 Ton light tanks at Camp Borden, October 1940.

471. The prototype Sentinel AC III mounting a 25pdr gun.

WITH war against Japan seeming more than probable, and with the added possibility even of a Japanese invasion, the Australian Ministry of Munitions first considered the idea of building tanks as early as July 1940. At this time, Britain's tank strength was inadequate for home defence, and there seemed little possibility of Australia receiving tanks from this source for some time to come. The Army Design Section (part of the Directorate of Mechanisation) was therefore asked to examine design characteristics and production problems, and in November 1940, the Australian General Staff drew up precise requirements for the sort of tank they thought necessary. They called for a 16-20 (long) tons vehicle, with 2pdr main armament, crew of 4-5, a range of 150 miles, and armour maximum of 50mm. They estimated that 2,000 would be needed, with first deliveries in July 1941 and output of 70 a week from then on.

The Ministry of Munitions asked the British General Staff for the services of a tank design expert from Britain, and, accordingly, a Colonel Watson was sent to Australia in December 1940. Watson travelled via America, where he had the chance to see the designs being drawn up for the M3 medium tank (qv), and on arrival in Australia he was appointed Director of Design. For the proposed vehicle, AC I (AC: Australian Cruiser), Watson planned to use a copy of the M3 final drive and gearbox since he had been impressed by the mechanical features of this vehicle. For a power plant, Guiberson diesel motors were planned but since it seemed probable that there would be difficulty in obtaining these, three commercial automobile engines, Ford at first, then more powerful Cadillac engines, were adopted, arranged in "clover leaf" formation. A leading Australian automobile engineer was co-opted to advise on development and installation.

In early 1941 a wooden mock-up of AC I was built. The vehicle was to have cast or rolled armour throughout, utilising only alloys available in Australia. By April 1941, drawings of the M3 final drive arrived from America, when it found that this installation was too sophisticated to be manufactured in Australia with existing facilities. Suitable machinery could not be delivered from Britain or America for at least another year. Meanwhile, the United States suggested that Australia produced a new design which could utilise components supplied from America. This proposal, envisaged the use of commercial truck engine and mechanical components. In July 1941, therefore, it was decided to go ahead with a new design which was designated AC II. The limitations which soon became evident using truck engines and drive, however, were many; principally the weight had to be kept below 16-18 tons with consequent reduction in armour thickness, and armament could be no heavier than a 2pdr gun. The truck mechanical components were not powerful enough to cope with a vehicle heavier than this. In September 1941, therefore, the AC II design was abandoned, and attention was given once more the the AC I.

It was found that by redesigning the final drive to a much more simplified form it would be possible to build the necessary components in Australia. Meanwhile, redesign had also been carried out on the bogies; originally vertical volute bogies of the M3 type were planned, but these were changed to horizontal volute pattern and proved much superior. The first cast hull was successfully manufactured in October 1941, and the prototype AC I was completed in January 1942. The hull and turret castings were in themselves a great achievement as nothing so complicated as this had previously been attempted by Australian industry.

Modifications were made to the prototype vehicle after trials, and in August 1942 the first production vehicle was completed at Chullona Tank Assembly Shops, NSW, only a year after the first over-optimistic (and unrealistic) estimate. Chullona Shops had been built starting in January 1942 specially to produce tanks, and were erected and managed by New South Wales State Railways, based on the American tank arsenals. A total of 66 AC Is were built when production ceased and all orders were cancelled in July 1943. By this time the tank supply situation had changed and the USA was able to provide all vehicles necessary for equipping the 1st Australian Armoured Division which had meanwhile been formed. The AC Is already completed were therefore used only for training and never saw combat service.

The Australian AC tank, named Sentinel, was a most remarkable achievement for a nation with only limited heavy engineering facilities and no previous experience of tank production. The arrangement of the Cadillac "clover leaf" power plant, and the cast one-piece hull were novel features which made a strong, tough, powerful vehicle capable of much future development. Plans for upgunned versions of the AC I (detailed below) never went beyond prototype stage, however, when AC production was prematurely terminated. Had AC manufacture continued, it was also planned to commence building ACs at the Geelong Tank Assembly Shops, Victoria, then being built, which were to be managed by Ford Motor Co (Australia).

SPECIFICATION:

Designation: Cruiser Tank AC I and AC III, Sentinel
Crew: 5 (commander, driver, hull gunner, gunner, loader) (No hull gunner in AC III).
Battle weight: 62,720lb
Dimensions: Length 20ft 9in; Height 8ft 5in; Width 9ft 1in; Track width 16½in; Track centres/tread 7ft 6½in
Armament: Main: 1 × 2pdr OQF (AC I); 1 × 25pdr (AC III)
Secondary: 2 × Vickers ·303 cal MG (one in AC III)
Armour thickness: Maximum 65mm; Minimum 25mm
Traverse: 360°. Elevation limits: —
Engine: 3 × Cadillac V-8 petrol, 117hp each unit (AC I)
Perrier-Cadillac triple engine (common crankcase), 397hp (AC III)
Maximum speed: 30mph
Maximum cross-country speed: 20mph (approx)
Suspension type: HVSS, Hotchkiss type
Road radius: 200 miles (AC I), 229 miles (AC III)
Fording depth: 4ft
Vertical obstacle: 2ft (AC I), 4ft (AC III)
Trench crossing: 9ft 6in
Ammunition stowage: 130 rounds 2pdr (AC I); 4,250 rounds ·303 cal (AC I)
Special features/remarks: Cast one-piece hull with prominent armoured sleeve for bow machine gun mount. HVSS copied from French Hotchkiss design in place of M3 type vertical volute suspension at first planned. Very low, stable, fast vehicle, with good armour protection and development potential. Tracks were American rubber block type. Bren AA machine gun mount fitted on cupola of all marks.

472. Standard production AC I Sentinel.

Variants

AC III: This was an upgunned design of the AC I mounting a 25pdr in place of the 2pdr. This necessitated considerable modification, mainly the provision of a larger turret and turret ring, which was increased from 54in to 64in diameter. The engine installation was redesigned with a common crankcase, allowing room for extra fuel tanks, and the bow machine gun was eliminated to give increased ammunition stowage. The bow machine gunner was also, of course, dispensed with, reducing the crew to four. A prototype for the AC III underwent trials in February 1943 and AC III production was to replace the AC I at Chullona from May 1943. However, in view of cancellation of the AC programme it seems probable that very few, if any, AC III were actually completed.

AC IV: The AC III prototype was subsequently tested in March 1943 with two 25pdrs in a co-axial mount, so that the feasibility of mounting the new 17pdr high velocity gun in the AC series could be investigated. Fired together, the two 25pdrs gave a recoil 20% greater than the recoil of a 17pdr gun with no adverse effect on the turret or vehicle. Plans thus went ahead to fit the 17pdr in the AC III design, and a prototype was completed and tested in late 1943. However, by then AC production had ceased, and no further production orders followed. With the 17pdr, the vehicle was designated AC IV. Undoubtedly it would have proved a most potent vehicle.

One AC hull was modified with torsion bar suspension in an attempt to provide superior riding qualities for the proposed upgunned models. Though completed and run, there was, of course, no opportunity of incorporating the new suspension in production vehicles.

473. AC III test vehicle with twin 25pdrs fitted.

474. AC IV prototype with 17pdr.

AUSTRALIAN VARIANTS

OTHER TANKS AND SP TYPES USED BY THE AUSTRALIAN ARMY, 1939-45

Vickers Medium Mk IIA: Four vehicles purchased from Britain in 1937 and modified by the addition of a ball-mounted ·303 cal Vickers MG in the turret to the right of the main armament. Still in use for training in 1939-42. For details see Medium Mk II, British section.

Vicker Light Tanks Mk VI, VIA, VIB: A small number of light tanks of these marks was obtained from Britain in 1939. They were used for training in Australia, mainly at Rokeby, the Australian mechanised cavalry centre, until 1942-43 when Australian-built Sentinels and ex-American vehicles replaced them. Australian forces also took over some vehicles of this type in the Western Desert in 1940, together with some Light Tanks Mk II, all ex-British vehicles. For details of these types see British light tank section.

Infantry Tank Mk II, Matilda III or IV: Vehicles delivered from British Army, 1942, after being superseded in British service. Used by Australians in New Guinea campaign.

Matilda Frog, Matilda Murray: Australian flame-thrower conversions. Frog was used in New Guinea campaign.

Matilda Dozer: Australian conversion of standard Matilda; used in New Guinea campaign.

For all Matilda details see British infantry tank section.

Covenanter Bridgelayer, Valentine Bridgelayer: Small number supplied to Australians. Mainly used for training. For details see British Covenanter and Valentine entries.

Infantry Tank Mk III, Valentine: Small number supplied to Australians from 1943 onwards. For details of this vehicle, see British section.

Medium Tank M3, Grant: Many of the Grants superseded by Shermans in the 8th Army in late 1942/early 1943, were supplied to the Australians in the Pacific theatre and used there until the war's end. For details see US M3 medium tank entry. The Australians also received a few M4 Shermans.

Grant ARV Mk II (Australian): Australian conversion of standard Grant tank to ARV role by removal of guns and installation of winch in fighting compartment, addition of tool boxes on hull rear, earth spade, and roller guide for winch purchase. Unique to Australian Army and distinct from British and US conversions of the Grant to the ARV (TRV) role.

Light Tank M3 series: Various models in this series supplied to Australians by the United States under Lend-Lease arrangements. For details of this type see US light tank section.

475. Light Tanks Mk VIA in use for training at Rokeby, Australia, October 1940.

476. Grant ARV Mk II (Australian); rear view showing spade and roller fittings.

477. Australian Army Marilda IVs patrolling the jungle at Finschafen, New Guinea, 1944. Nearest vehicle is a CS model. Markings have been obliterated by the censor in this picture.

478. Schofield wheel-and-track tank showing vehicle rigged for running on wheels.

IN 1940, New Zealand was in very much the same position as Australia, faced with the possibility of eventual involvement in a war with Japan and the further possibility of invasion. Since no tanks could be procured from Britain at that critical period, the New Zealand government was forced into considering building tanks of its own. The only true tank which actually reached prototype stage was designed by E. J. Schofield of General Motors (Wellington) and was based on the chassis of the GMC 6cwt commercial truck which was built in New Zealand. Track and suspension units were taken from a Universal Carrier, which was just going into local production in New Zealand, while the wheels and mechanical units came from the 6cwt truck. Armour plate for the hull and turret was produced by New Zealand Railways at Hull Valley Works. The cylindrical turret was open-topped and mounted a 2pdr gun and co-axial machine gun. The truck wheels on this vehicle shared common stub axles with the sprockets and idlers, and the change from wheels to tracks was effected by pivoted arms operated from the hull interior. When running on wheels, the tracks had minimal ground clearance and were kept clear of the ground by chains. When rigged for running on the tracks, the wheels were removed and fitted to stub carriers on the hull sides. The Schofield tank was completed in August 1940, but was apparently impractical since no production order followed. In mid 1943 the prototype was sent to Britain for trials.

SPECIFICATION:

Designation: Light Tank, Wheel-and-Track (Schofield)
Crew: 3 Duties: Commander/Gunner/Driver
Battle weight: 11,680lb
Dimensions: Length 13ft 1in; Height 6ft 7½in (6ft 10½in on wheels); Width 8ft 6½in; Track width 9½in; Track centres/tread —
Armament: Main: 1 × 2pdr OQF
Secondary: 1 × 7·92 cal Besa MG (co-axial)
Armour thickness: Maximum 10mm
Minimum 6mm
Traverse: 360°. Elevation limits: —
Engine: Chevrolet 6 cylinder petrol, commercial type. 29·5hp.
Maximum speed: 25·7mph (tracks), 45·6mph (wheels)
Maximum cross-country speed: —
Suspension type: Hortsmann
Road radius: —
Fording depth: —
Vertical obstacle: —
Trench crossing: —
Ammunition stowage: —
Special features/remarks: Ingenious design of a type proved largely impractical by British experiments in the inter-war years.

OTHER TANKS AND SP TYPES USED BY THE NEW ZEALAND ARMY, 1939-45

"Bob Semple" Tank: This was an idea to produce an armoured superstructure for fitting to an International Harvester agricultural tracked tractor chassis. It consisted of a box-like structure with loopholes for rifle fire from the troops carried inside. The vehicle weighed about 20-25 (long) tons and had a top speed of 8mph. The name was derived from that of the New Zealand defence minister who instigated their production. Only four were built in 1940-41, but they were severely impractical, top heavy, and unstable due to the limited resilience of the suspension of the tractor

479. Rear view of the Schofield wheel-and-track Tank, shown with wheels down for road running.

480. New Zealand "Mobile Pillbox" tank, an improvised superstructure on an agricultural tractor which was also known as the "Bob Semple" tank.

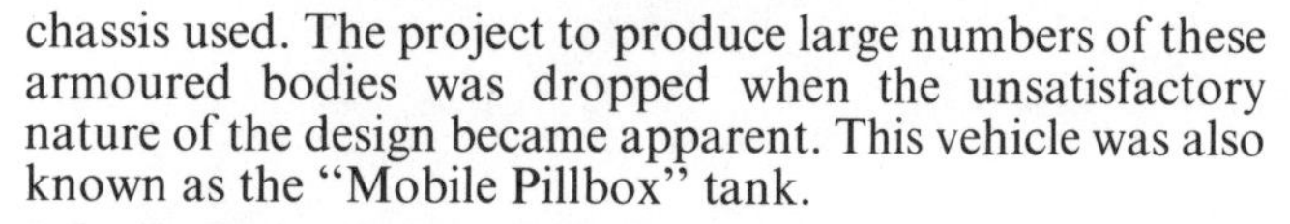

chassis used. The project to produce large numbers of these armoured bodies was dropped when the unsatisfactory nature of the design became apparent. This vehicle was also known as the "Mobile Pillbox" tank.

Infantry Tank Mk III, Valentine III: A number of these were delivered to New Zealand in 1942 and were used by New Zealand forces in the SW Pacific theatre. Some were converted to CS tanks by replacing the 2pdr guns with 3in howitzers taken from Australian Matilda tanks converted to flame-throwers. For Valentine details see British infantry tank section. New Zealand also used Matilda, Shermans, and Stuart (M3 light) tanks in World War II. For details of these vehicles, see relevant entries in British and American sections.

481. Stuart I (M3) light tanks, seen on training exercises in New Zealand, 1943.

PART 4

SELF-PROPELLED GUNS ON HALF-TRACKS AND CARRIERS

482. A 75mm Gun Motor Carriage M3 of the US Marines comes ashore from a LST during the landings at Cape Gloucester, December 1942. This vehicle carries a ·50 cal MG on each of the pintles provided for the purpose.

HALF-TRACK VEHICLES USED BY ARMOURED UNITS

483. A heavily laden Half-track M3A1 towing a 105mm howitzer passes through Zweibrücken during the US 7th Army advance into Germany, March 1945. Note the pulpit with the ring mount for the ·50 cal MG.

UNITED States Army interest in the half-track dated back to 1925 when the Ordnance Department purchased two Citroen-Kegresse semi-track vehicles from France. They bought another in 1931. US commercial firms undertook development work on half-tracks on behalf of the Ordnance Department and the first indigenous design, the T1 Half-track, was built by Cunningham of Rochester, NY, in 1932. The development story of these vehicles in the thirties is beyond the scope of this book, but by 1939-40 Half-track Personnel Carrier T14 had been produced and became the prototype of all subsequent half-track types used by the US in World War II. In September 1940 the T14 was standardised as the Half-track M2 and, with modifications in order to transport personnel, it was standardised as the Half-track Personnel Carrier M3.

The M2 and M3 were similar in design and all major assemblies were interchangeable. The chassis and drive units were basically commercial components. The armoured hull was ¼in thick and included armoured shutters over the radiator, while armoured shields (½in thick) were provided for the cab windscreen and side windows. Vehicles were built with either an unditching roller mounted ahead of the front fender (though this was sometimes removed) or else with a winch. Late production vehicles also had stowage racks on the hull sides.

The M2 was basically a gun tower with the appropriate ammunition stowage facilities, and the M3 was a personnel carrier with slightly longer hull and rearranged seating. Contracts for production of these vehicles went to White and Autocar (M2) and Diamond T (M3) in September 1940. White had produced the Scout Car M2A1, one of which had been converted to a half-track during development work leading to the T14 prototype. Thus the Scout Car (not covered in this book) and the half-track had similar superstructures. The large orders placed by the US Army for half-track vehicles in the autumn of 1940 made standardisation of components essential among the contractors, and White, Autocar, and Diamond T agreed to such a policy under the overall supervision of the Ordnance Department.

First production M2s were delivered in May 1941 and a total of 11,415 were built, 2,992 by Autocar and 8,423 by White. Total M3 half-track production reached 12,499 units. For AA protection, the M2 had a track running round the inner edge of the superstructure, on which was mounted skates for ·30 and ·50 calibre MGs, while the M3 had a pedestal mount instead. Modifications to this arrangement in 1943 led to the elimination of these types of gun mounts in favour of a "pulpit" with circular gun mount just behind the cab. Three pintles were provided, also for AA machine guns. With these new features the vehicles were redesignated M2A1 and M3A1 respectively. These replaced the earlier models in production from October 1943 and 2,862 M3A1 and 1,643 M2A1s had been built when production ceased in March 1944. A White engine was used in all these models.

Further variants based on the M2A1, but designated separately, were built as mortar carriers. In 1942, also, when there was an increased requirement for half-tracks, International Harvester were brought into the half-track production group and built models with minor detail improvements and International, instead of White, engines. Designated Half-tracks M9 and M5, these corresponded to the M2 and M3 respectively. With the later production improvements came the M9A1 and M5A1 which corresponded to the M2A1 and M3A1 produced by the White/Autocar/Diamond T plants.

In April 1943, work started on rationalising the half-track design to produce a "universal" vehicle with common body features suitable for either the gun tower/mortar carrier role or the personnel carrier role. This led to the M3A2 and M5A2 types from White/Autocar/Diamond T and International Harvester Co respectively. Standardisation of these revised designs took place in October 1943 but production was later cancelled. By this time, in fact, US Army interest in the half-track was beginning to wane and production of this type of vehicle tailed off completely in mid 1944, though half-tracks remained in wide service with the American forces until the war's end. For artillery use the half-track was being displaced as a gun tower by the increasing availability of the high speed full-track tractor and in other service arms there was a growing preference for either full-track utility vehicles or trucks for personnel and supply work. In fact, half-tracks were never fully replaced in the period covered by this book though the process had started in 1944. Total US half-track production reached 41,169 units.

While the half-track was initially conceived as a fast reconnaissance vehicle, protected against small arms fire and with a good cross-country ability mainly for infantry and artillery use, it was also widely employed by other arms including the Armored Force as a "utility" vehicle. In this respect it was roughly equivalent in the US Army to the British Universal Carrier, but its larger size gave it more development potential as a weapons carriage. For the Armored Force, the Ordnance Department produced a number of expedient designs of gun motor carriage on the basis of the half-track and these performed useful "stop-gap" service while superior full-track motor carriage designs were perfected. British armoured units also used a number of these half-track motor carriage types, supplied under Lend-Lease arrangements.

There were scores of special purpose variants based on the US half-tracks but many of these were not used by armoured units and strictly speaking fall more properly into the armoured car category not covered in this present volume. In this section, therefore, only those half-track variants used by, or specially developed for, armoured units are described and illustrated.

Basic Models

Half-track Car M2: This was the basic vehicle intended for use as an artillery prime mover mainly for the 105mm howitzer (on field carriage) in artillery battalions. It had seats for 10 men (ie, a full gun crew) plus seats in the cab for the driver, assistant driver, and commander. There was no rear door and the body was shorter than that of the M3 (see below). There were two ammunition lockers sited on each side of the interior immediately behind the driving cab with access doors to these from outside the vehicle but with opening lids giving access from inside to the top shelf only. A skate rail for AA machine guns was fitted round the inner edge of the superstructure and one ·50 cal and one ·30 cal Browning MG were fitted to tracks moving on the skate rail. An unditching roller or winch was fitted at the front end and the vehicle was radio equipped.

Half-track Car M2A1: This vehicle replaced the M2 in production and differed in having an M49 ring mount for a ·50 cal MG sited in a "pulpit" over the co-driver's seat. The skate rail was eliminated. Three pintles were fitted, two at the side and one at the rear of the superstructure for optional fitting of one ·30 cal MG. Prototype vehicle for this development was designated M2E6 and this vehicle, appropriately modified, also became the prototype for later production variants of the basic half-track.

Half-track Personnel Carrier M3: Produced concurrently with the M2, this was the basic personnel carrier model. It had seats for 10 men (ie, a rifle squad or section) in the rear and seats for three in the cab, as in the M2. The body was about ten inches longer than that of the M2 and had an access door in the rear. Due to this, there was no skate rail in the M3 and a pedestal mount M25 was provided for a ·30 cal MG, this being fitted to the floor of the rear compartment and demountable as required. The vehicle was produced fitted with either an unditching roller or winch at the front end.

484. The basic Half-track M2 with ·50 and ·30 cal MGs fitted.

485. Basic Half-track M3 with ·50 cal MG on pedestal mount.

486. 81mm Mortar Carrier M4 showing location of skate rail for AA machine guns.

487. 81mm Mortar Carrier M4A1 showing internal layout and crew positions.

HALF-TRACKS

Half-track Personnel Carrier M3A1: This was the modified M3 design with ring mount M49 and "pulpit" for AA machine gun, equivalent to the M2A1.

The M3 and M3A1 models were widely used by armoured units as ambulances, command cars, and general utility and liaison vehicles. In British service, some were converted to recovery vehicles by the addition of a front-mounted jib.

Half-track Car M3A2(T29): Based on the M3 half-track, this was a vehicle for "universal" use to replace both the M2 and M2A1 and the M3 and M3A1 series. International Harvester Co converted an M3 as a prototype with interior fittings suitable for the gun tower or personnel carrier role. The prototype, designated T29, was completed by July 1943 and standardised as the M3A2 in October 1943. Though production plans were made, no production took place due to curtailment of the half-track programme. In appearance, this vehicle resembled the M3A1 but had an armoured shield on the AA ring mount and movable stowage boxes in the rear hull which could be changed according to the role required for the vehicle. Seats for from five to twelve men could be fitted.

Half-track Personnel Carrier M5: Built by International Harvester, this vehicle corresponded to half-track M3 but featured International Harvester components, including an IHC 6 cylinder engine instead of the White engine of the M2/M3 series. The body was of homogenous plate ($\frac{5}{16}$in thick) instead of face-hardened plate as in the M2/M3 series. Major external distinguishing features resulting from this were the rounded rear corners of the superstructure. Additionally, the mudguards on the M5 were flat instead of rounded. Other physical characteristics were similar to those of the M3. Due to their "non-standard" nature, many of the M5 half-tracks were allocated to Lend-Lease stocks and most went to the British Army.

Half-track Personnel Carrier M5A1: This was the M5 modified with M49 ring mount to correspond with the M3A1.

Half-track Car M5A2(T31): This was a projected vehicle, corresponding to the M3A2 (qv, above) for "universal" use. International Harvester modified one M5 half-track under the designation T31 and this was standardised and ordered into production in October 1943. All production vehicles were specifically ear-marked for Lend-Lease allocation. However, following curtailment of half-track requirements, all production was cancelled early in 1944.

Half-track Car M9A1: This was the half-track variant produced by International Harvester corresponding to the M2A1 for the gun tower role. However, unlike the M2A1 half-track, it had a rear access door and body length matching the M5. There was no M9, since designation of the proposed M9 was changed to M9A1 (with corresponding alterations) before production began. 3,433 of these were built.

81mm Mortar Carrier M4 and M4A1: The purpose-built mortar carrier version of the M2 half-track, this vehicle was fitted to accommodate crew, mortar and ammunition. Seating capacity in the rear was reduced to three, the remaining space being taken up by ammunition racks and stowage for the mortar. In the M4 there was no provision for firing the mortar from the vehicle, though this could be done in emergency situations. A total of 572 M4s were built by early 1942 when the model was replaced by an improved design, M4A1, which had a reinforced floor and mounts for firing from the vehicle. White built 600 M4A1 mortar carriers in 1943. These vehicles could be distinguished from

488. 81mm Mortar Carrier M21 showing forward firing mortar and ·50 cal MG on pedestal mount.

489. 75mm Gun Motor Carriage M3, most successful of the motor carriage adaptations on the half-track chassis.

490. 75mm Gun Motor Carriage T73 was a later prototype based on the M3 (Plate 489) but with the 75mm Gun M3 instead of the M1897A model.

491. 75mm Howitzer Motor Carriage T30 was similar in layout to the M3 GMC but had howitzer armament.

492. 75mm Gun Motor Carriage M3 in service with a British armoured regiment gives support fire on the 8th Army front, Italy, March 1945.

the earlier M4 models by the addition of extra stowage boxes at the rear.

81mm Mortar Carrier M21(T19): The limitations of the M4 and M4A1 mortar carriages, which included the rear facing firing position and lack of traverse for the mortar, led to the development of a much improved design based on the M3 half-track. Major change was the provision of a traversing arc and reinforced mount for firing forward from within the vehicle. This gave a traverse of 30° each side and an elevation range of 40°–80°. The hull was also longer and roomier than in the earlier models based on the M2 half-track. A pedestal mount was provided in the hull rear for a ·50 cal MG. The standard M1 81mm Mortar carried could also be demounted for firing from the ground. The prototype vehicle, designated T19, was standardised in June 1943 and 110 vehicles were built.

4·2in Mortar Carrier T21 and T21E1: Similar in layout to the M4 and M4A1 mortar carriers, development of this vehicle was initiated in December 1942 using the M3 half-track as a basis. The mortar was mounted to fire to the rear as in the M4A1. Modifications were requested, however, to bring the layout into line with the improved arrangements of the M21 mortar carrier. As so modified with forward firing mortar on traversing arc, the vehicle was designated T21E1. Interest switched to full-track chassis, however, (see Mortar Motor Carriage T27) and the T21E1 project was dropped.

Gun and Howitzer Motor Carriages on Half-Track Chassis

While the half-track served as the basis for numerous self-propelled weapons, relatively few designs were standardised or put into service and those which were were generally regarded as expedient designs while full-track motor carriages were developed. Notable exceptions were the standardised AA vehicles which remained in service until the end of the war. Due to the general shortage of full-track chassis at the time of rearmament in 1941 the Ordnance Department developed several SP weapons on the M3 chassis and these are described first.

75mm Gun Motor Carriage M3(T12) and M3A1: The urgent need for a tank destroyer to be rushed quickly into service led to the adaptation of the M3 half-track in June 1941 to take a suitably modified M1897A 75mm gun on a pedestal mount firing forward. The M1897A was the American version of the famous French "75", dating from World War I, of which surplus stocks were available. Designated T12 GMC, the gun had a limited traverse and was provided with a shield. Despite its extemporary nature, this equipment proved most successful on trials and the vehicle was standardised as the M3 GMC in October 1941. First vehicles built were sent instantly to the Phillipines at the end of 1941 in time to see action against the Japanese. Aside from wide use in the Pacific, M3 GMCs were also used by US forces in the North African (Tunis) campaign and on replacement by full-track tank destroyers, these vehicles were handed over to the British who used them (in Italy) until the end of the war. In British service the M3 GMC was known as the "75mm SP, Autocar" (Autocar being the builder of this model), and the vehicles were used in HQ troops of armoured car and tank squadrons to give support fire.

Due to shortage of the original 75mm gun mount, later vehicles were produced with a modified mount under the designation 75mm Gun Motor Carriage M3A1. They were externally similar to the M3 GMC.

75mm Howitzer Motor Carriage T30: Following the success of the M3 GMC (above), the Ordnance Department carried out a similar conversion of the M3 half-track to mount a 75mm howitzer. Two pilot models were authorised in January 1942 and an order for 500 vehicles was placed following successful firing trials. The 75mm Howitzer M1A1 carried in this vehicle had a maximum elevation of 25° and a traverse 21° left and 23° right. Though never standardised,

493. 57mm Gun Motor Carriage T48 which was supplied only to the British.

494. 105mm Howitzer Motor Carriage T19. Some of these vehicles had gun shields.

495. Multiple Gun Motor Carriage T28 was prototype for GMC M15. This view shows it with the 37mm gun before the addition of the two ·50 cal MGs on the same mount and before the fitting of the gun shield.

496. Multiple Gun Motor Carriage T1E4, standardised as the M13, was the first production AA variant on the half-track chassis. Note the folding superstructure sides.

the T30 equipped the HQ companies of the medium tank battalions until replaced by the 75mm Howitzer Motor Carriage M8 (qv) in 1943. It was regarded only as a "stop-gap" type.

105mm Howitzer Motor Carriage T19: Another type produced following experience with the M3 GMC, this was the largest calibre SP type on the half-track chassis. Similar in layout to the T12 and T30 it incorporated the 105mm Howitzer M2. A total of 324 were built, some with and some without gun shields. They were used in action in North Africa (Tunis) until replaced by the M7 Howitzer Motor Carriage (qv). The type was never standardised.

105mm Howitzer Motor Carriage T38: This was a project only to mount a short barrel 105mm Howitzer on the M3 half-track in similar style to the T19.

75mm Gun Motor Carriage T73: A modified version of the M3 GMC but with a modern 75mm Gun M3 replacing the M1897A weapon.

57mm Gun Motor Carriage T48: Development of this vehicle started in April 1942 at APG to meet British and American requirements for a self-propelled anti-tank gun. It was an adaptation on the M3 half-track similar to the earlier designs, and incorporated the 57mm M1 gun, an American-built version of the British 6pdr. A shield with overhead plate was fitted to the gun. In October 1942 the American requirement for the T48 was dropped but development continued at British request. A total of 962 vehicles were ordered of which 680 were delivered to Britain on Lend-Lease. The remainder were retained in the United States and converted to M3A1 personnel carriers. White built these T48 vehicles in 1942-43. Most of the vehicles delivered to Britain appear to have been converted to personnel carriers by removal of the gun. The T48 had British (No. 19 set) radio.

Multiple Gun Motor Carriage T1E1: This was an early project to mount a Bendix power-operated turret with twin ·50 cal MGs on a M2 half-track in a series of comparative trial installations, 1941, to evaluate a design of motor carriage for AA defence in the field.

Multiple Gun Motor Carriage T1E2: The T1E2 was similar to the T1E1 except that a Maxson twin ·50 cal MG turret was fitted instead of the Bendix turret. It was tested against the T1E1.

Multiple Gun Motor Carriage T1E3: This was a further test project like the T1E1 and T1E2 except that an Electro-Dynamic power-operated turret of aircraft type was fitted. Of the three installations, that of the T1E2 was preferred.

Multiple Gun Motor Carriage T1E4(M13): For the production model using the Maxson turret, it was decided to use the M3 half-track as a basis. Otherwise similar to the T1E2, the T1E4 offered superior stowage space due to the longer body. The Maxson turret had 360° traverse, maximum elevation of 90° and depression of —11·5°. The turret could traverse in power operation at 74° a second. Standardised as the Multiple Gun Motor Carriage M13, a total of 535 were built by White in 1942.

Multiple Gun Motor Carriage M14: To speed up production, the same Maxson gun mount (designated M33) was installed in the M5 half-track under the designation Multiple GMC M14. It was similar to the M13 except for the different mechanical and physical characteristics of the M5 half-track series (qv). A total of 1,905 units of this type were built by International Harvester Co in 1943. Many M14s were supplied to Britain under Lend-Lease but these were modified for the personnel carrier role by the British who removed the gun mounts.

Multiple Gun Motor Carriage T28 and M15/M15A1 (T28E1): To give increased fire power a new design was evolved, the T28, based on the M2 half-track and consisting of the top part of the 37mm Gun Carriage M3E1 combined with two ·50 cal MGs on a combined traversing mount with armoured shield. The project was dropped in favour of the same installation on the M3 half-track designated T28E1. With minor changes resulting from trials this vehicle was standardised as the GMC M15 and 680 were built by Autocar in 1942. They were first used, with great success, by American forces in the North African (Tunis) campaign. With minor modifications, including an altered gun mount (see plates 500 and 502), later vehicles produced from August 1943 were designated M15A1.

Multiple Gun Motor Carriage M16 and M17(T58): In April 1942 development work started on a new Maxson turret with four instead of the two ·50 cal machine guns of the M33 mount on the M13/14. The new turret was tested on a M2 half-track chassis, and later on a M3 half-track chassis under the designation GMC T58. The new mount was standardised as the M45 and the T58 vehicle with this mount

497. First prototype AA vehicle on a half-track chassis was the T1E1 with Bendix power-operated turret.

498. Comparative test vehicle in trials with the T1E1 (plate 497) was the T1E3 with aircraft type power-operated turret.

499. Multiple Gun Motor Carriage T61 was a prototype with improved Maxson quad turret. This vehicle was converted from the original T1E2 AA vehicle, the designation of which can be seen still painted on the cab side.

500. Multiple Gun Motor Carriage M15A1 had a 37mm cannon and two ·50 cal MGs in a combination mount. For prototype vehicle T28 and mount details see plate 495.

501. Multiple Gun Motor Carriage T10 was another prototype which did not go into production. It mounts a twin Maxson turret with 20mm cannon.

502. Detail view of the Multiple Gun Motor Carriage M15 showing crew disposition. In the M15 the ·50 cal MGs were placed above the 37mm cannon and in the M15A1 (plate 500) they were below it.

503. 40mm Gun Motor Carriage T59E1 was a subsequent development to the T54, with outriggers to give improved stability.

504. Most complex half-track AA carriage was the T68 which had twin superimposed 40mm guns. It was experimental only.

was standardised as the M16. A total of 724 vehicles of this type were built by White in 1942-43. The same installation based on the M5 half-track (qv) was designated M17 and was of similar external appearance. International Harvester built 1,000 of these in 1943-early 1944 when all half-track production was terminated.

Multiple Gun Motor Carriage T37 and T37E1: These were two projects to investigate improved quad ·50 cal MG mountings for AA defence. The T37 was a M3 half-track carrying a T60 mount in which the four ·50 cal MGs were arranged in box formation (2×2) and the T37E1 was the same vehicle with mount altered (as the T60E1) so that the four guns were in line. A circular gun shield (½in thick) was provided. No production order followed.

40mm Gun Motor Carriage T54 and T54E1: Development of a half-track motor carriage to mount a 40mm AA weapon began in June 1942 when a M3 half-track was adapted to carry a single 40mm Gun Mount M1 (Bofors). Pilot model, designated T54, showed the need for improved stability and changes were incorporated in a second pilot model, designated T54E1. It was finally decided that outriggers and jacks would be necessary if the vehicle was to provide a satisfactory gun platform, and the project was ended in favour of improved designs (see below).

40mm Gun Motor Carriage T59 and T59E1: The T59 was based on the T54E1 but incorporated folding outriggers and jacks. The T59E1 was the same vehicle modified to take the power-operated Firing System T17. To go with this system a companion vehicle was developed, the Half-track Instrument Carrier T18 which was a basic M3 carrying a M5 Director and M5 Generating Unit to provide power and control. No production took place of either of these vehicles, however, since a full-track AA vehicle (the M19 GMC, qv) was developed on the M24 light tank chassis and had much superior characteristics.

Multiple Gun Motor Carriage T60 and T60E1: This was an adaptation of the T54 and T59 designs to provide a vehicle with a combination gun mount (designated T65) which mounted the 40mm Gun M1 and two ·50 cal MGs in similar fashion to the M15 GMC (qv). With modified shield and stowage the vehicle was designated T60E1. No production orders were placed.

40mm Gun Motor Carriage T68: This was an experimental vehicle based on the M3 half-track which had twin superimposed 40mm guns with overhead equilibrators. Though a pilot model was built, no further progress was made with this design.

Multiple Gun Motor Carriage T61: An experimental mounting of a quad ·50 cal MG on a M2 half-track chassis. Not proceeded with.

Multiple Gun Motor Carriage T10 and 20mm Gun Motor Carriage T10E1: In July 1941, when work was in progress on developing half-track motor carriages with AA machine guns (see T1 GMC series above), experiments were also conducted to investigate the feasibility of mounting heavier weapons. Guns considered were the 20mm Oerlikon, Hispano-Suiza 20mm, or 20mm Automatic Guns AN-M1 or AN-M2. These were to be fitted in an aircraft type power-operated turret built by Maxson. Fitted to a modified M3 half-track chassis, the vehicle was designated T10. A modified gun mount fitted later to a modified M16 GMC chassis was designated T10E1. No production orders were placed.

505. Multiple Gun Motor carriage M14 was similar to the M13 (plate 496) but was based on the M5 half-track chassis instead of the M3 chassis. Note the rounded rear corners of the superstructure and the flat section mudguards, characteristic of the M5 half-track chassis. Compare with the rounded mudguards and squared off rear corners of the M3 superstructure in plate 496. Note the spare magazines in the stowage behind the driving compartment.

506. Multiple Gun Motor Carriage M16 was an improved model with four, instead of two, ·50 cal MGs in a Maxson turret. Mount was designated M45. Note cut-away section in folding side plates to clear magazines. This is a winch-fitted vehicle.

SPECIFICATION:

Designation: Carrier, Personnel, Half-track M3 or M3A1
Crew: 3 plus 10. Duties: commander, driver, co-driver plus 10 passengers.
Battle weight: 20,000lb
Dimensions: Length: 20ft 2⅝in Track centres/tread: 5ft 3¼in
Height: 7ft 5in
Width: 7ft 3½in
Armament: 1×·50 cal MG/1×·30 cal MG
Armour thickness: Maximum 12·72mm
Minimum 6·35mm
Engine: White 160 AX, 147hp
Maximum speed: 45mph
Suspension type: Vertical volute springs (rear tracks).
Road radius: 180-215 miles
Fording depth: 2ft 8in
Vertical obstacle: 1ft
Ammunition stowage: 700 rounds ·50 cal
7,750 rounds ·30 cal
Special features/remarks: Details given for basic M3/M3A1 only. SP motor carriage variants differed in weight, height, ammunition etc, according to armament. M5, M5A1, M9A1, and SP variants on these chassis were built by International Harvester Co and had International RED–450 (143hp) engines and slightly thicker armour and increased weights.

507. The AA variants of the US half-track saw wide use with US forces during the NW Europe campaign. Here a Multiple Gun Motor Carriage M16 is emplaced to guard Ludendorff Bridge at Remagen (captured intact) as troops of the US 1st Army move across into the bridge-head on the east bank during the crucial Rhine crossing. Another M16 moves into position in the background. This picture was taken on March 11 1945, four days after the initial Rhine crossing. Despite Allied air superiority at this period, the Luftwaffe still contrived to attack key targets like this bridge.

508. 40mm Gun Motor Carriage T54 was a project to mount the Bofors gun on the half-track M3 chassis.

MISCELLANEOUS

Half-track Car M2 with Mine Exploder: One M2 half-track was experimentally modified as a mine clearing vehicle with a light girder structure and flail carried ahead of the bonnet. No further progress was made with this after trials.

Other experimental work on half-tracks included an armoured roof fitted on one M2 and another M2 (designated M2E4) fitted with a Hercules engine. A series of prototypes, designated Half-track Trucks T16, T17, T19 respectively were produced by Diamond T, Autocar, and Mack respectively in 1943 as proposed replacement vehicles for the M2/M3/M5 series half-tracks. They featured longer tracks and many improvements. However, curtailment of half-track production in early 1944 led to abandonment of these projects. They had been designed for "universal" use as gun carriages, gun towers, APCs, etc.

SELF-PROPELLED WEAPONS ON INFANTRY CARRIER CHASSIS

THOUGH developed essentially for infantry use, the Universal Carrier and Loyd Carrier (and equivalent vehicles produced in Canada, Australia, and New Zealand) were also employed by other arms for various roles. Armoured units and SP artillery regiments used them for liaison, OP, and ambulance duties. Development of the Universal and Loyd Carrier types is outside the scope of this book, but various miscellaneous self-propelled weapons and other types based on Carrier chassis are included in this section for completeness. Most were of an experimental or expedient nature. Few, if any, of the SP types proved practical due to the small size of the Carrier chassis which imposed severe limitations on the size of the weapon which could be mounted.

509. Loyd 2pdr SP anti-tank gun.

(1) British Types

In 1940 when there was a severe shortage of armoured vehicles of all kinds, REME Workshops devised and constructed several SP vehicles on the Loyd Carrier chassis. One of these (plate 510) had a 2pdr anti-tank gun mounted centrally in the hull with all-round traverse. A second vehicle (plate 509) had the 2pdr gun mounted low in the hull front with a fabricated shield. Traverse was limited to 270° in this vehicle. Both were loosely designated "Loyd 2pdr SP A/T gun". A later conversion of early 1941 had a 25pdr gun/howitzer mounted in the front hull of a Loyd Carrier (plate 511). An experimental AA conversion on the Loyd Carrier featured a centrally mounted traversing platform with an armoured covered seat for the gunner and a quad Bren LMG installation with associated sighting equipment. An armoured cover was fitted for the driver. This was another 1941 expedient type. None of these prototypes saw service.

An interesting conversion by the British Home Guard which was used in small numbers in 1940-41 for home defence (though never tested under combat conditions) was the Smith Gun on a Bren Carrier. The Smith Gun was a primitive smooth bore 3in weapon, more exactly a projector, one of several extemporised types evolved for Home Guard use. A few were mounted in Carriers with suitably modified superstructure forming a very crude and primitive type of SP gun (plate 512). A more official conversion on the Universal Carrier chassis was the "Praying Mantis" (plate 515). This had two remote controlled Bren guns in a hydraulically operated elevator which could be raised 11 feet or lowered flat on to the carrier as necessary. The gunner sat inside the elevator and the driver sat centrally in the chassis. The idea was that the guns could be raised to fire over hedgerows, then lowered again while the vehicle made good its escape under cover. It was another anti-invasion device, a prototype only.

At the same time as various light tanks were being converted to AA tanks, a Universal Carrier was similarly modified as a prototype (plate 517). A turret with twin Vickers "K" machine guns and sights was fitted over the left gun sponson, the shape of which was suitably altered. The driver's compartment on the right was similarly plated in. Though this vehicle was tested, no production order appears to have been given.

An important conversion of the Universal Carrier widely used by armoured units was the Carrier Ambulance (plate 522) which was a Universal Carrier modified with lengthened superstructure and open rear end to allow a stretcher to be carried each side of the engine casing. This vehicle was used for evacuating injured tank crewmen under fire.

510. Loyd Carrier with 2pdr anti-tank gun.

511. Loyd Carrier with 25pdr gun/howitzer.

512. Bren Carrier with Smith Gun (Home Guard service).

513. Australian Carrier 2pdr Tank Attack. Note lengthened chassis.

514. Australian Carrier 3in Mortar. (Figure 1 indicates baseplate for mortar used for firing from ground only and carried on side of engine compartment).

515. Praying Mantis.

516. Loyd Carrier AA with quad Bren mount. Note armour protection for both gunner and driver.

(2) Canadian Types

Canada built 33,987 Universal Carriers during World War II, many of which were supplied to Britain. There were only two variants which came in the true SP gun category, however, the Universal Carrier Mk I* and II* with a 2pdr anti-tank gun fitted on the main transverse bulkhead (plate 519), and the Universal Carrier fitted with a PIAT battery, an extemporised SP type produced by a Canadian unit in NW Europe in 1944 (plate 518). This had 14 PIAT (Projectile Infantry, Anti-Tank) projectors mounted on a frame at the rear of a standard Universal Carrier and was used to give support fire in a similar fashion to Rocket Launchers mounted on some US medium tanks.

The vehicle with the 2pdr was officially designated **Carrier, Universal, 2pdr Equipped.** Ammunition was stowed sides and front of the division plate, and the engine cover was re-designed to give clearance for the gun recoil.

(3) Australian Types

The Universal Carrier built in Australia differed in several respects from the British model. It was the only AFV built in quantity in Australia during World War II and was an all-welded vehicle (the British Universal Carrier was riveted) with extra stowage boxes at the rear and better ventilation for the engine. In 1940 an experimental version was produced as a prototype with a 2pdr anti-tank gun mounted behind the left sponson, the top of which was cut down to clear the gun mount (plate 521). This vehicle was used for training. The standard Universal Carrier was also adapted as a 3in mortar carrier (plate 520) with reinforced engine cover allowing the mortar to be fired from the vehicle. The vehicle with the 2pdr gun featured a hydraulic lift to raise the gun to its firing position.

These adaptations led to the design of a much modified purpose-built chassis which was used as the basis for a 2pdr SP and a mortar carrier. Designated **Carrier 2pdr Tank Attack** (plate 513), the former had a slightly lengthened chassis and the engine was moved from the centre to the left front. The resulting clear rear deck was used to mount a complete 2pdr anti-tank gun less its wheeled carriage. The gun had full 360° traverse. Details: Crew: 4. (Driver and three gunners), Length: 13ft 3in, Width: 6ft 7in, Height: 6ft 2in, Battle Weight: 11,200lb, Armament: 1 × 2pdr and

517. Universal Carrier, AA, prototype vehicle.

518. Canadian Universal Carrier with PIAT battery fitted, 1944.

521. Australian Universal Carrier with 2pdr anti-tank gun.

519. Carrier, Universal, 2pdr Equipped. Used only for training in Canada.

520. Australian Universal Carrier adapted as 3in mortar carrier. Note all-welded construction and prominent engine vent.

1 × Bren LMG, Ammunition: 80 rounds 2pdr in 10 racks on gunshield, Engine: Ford V8, 95hp, Max speed: 20mph.

The mortar carrier version was designated **Carrier 3in Mortar** (plate 514) and had the same chassis modifications as the 2pdr Tank Attack Carrier, but a protected compartment was built at the rear carrying crew, mortar and ammunition. There was a turntable on the floor of the rear compartment allowing the mortar to be traversed 360° for firing from within the carrier. It could also be demounted and fired from the ground. This was a most handy adaptation of the basic Universal Carrier idea and a considerable advance on the original design. All details (except ammunition and armament) were the same as for the 2pdr Tank Attack Carrier.

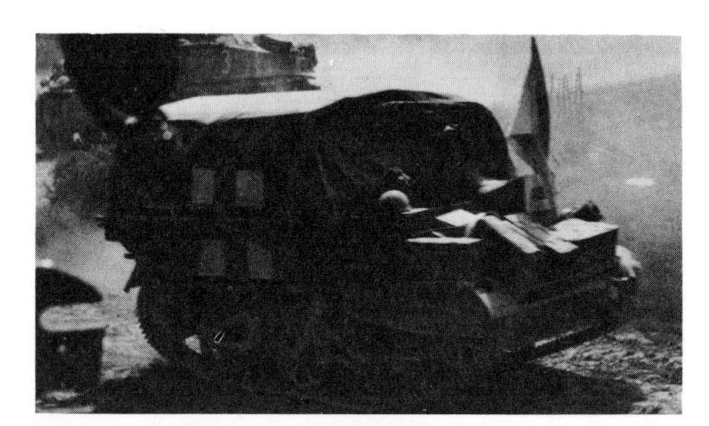

522. Carrier Ambulance moving forward with Sherman tanks, Italy, 1945.

PART 5

APPENDICES

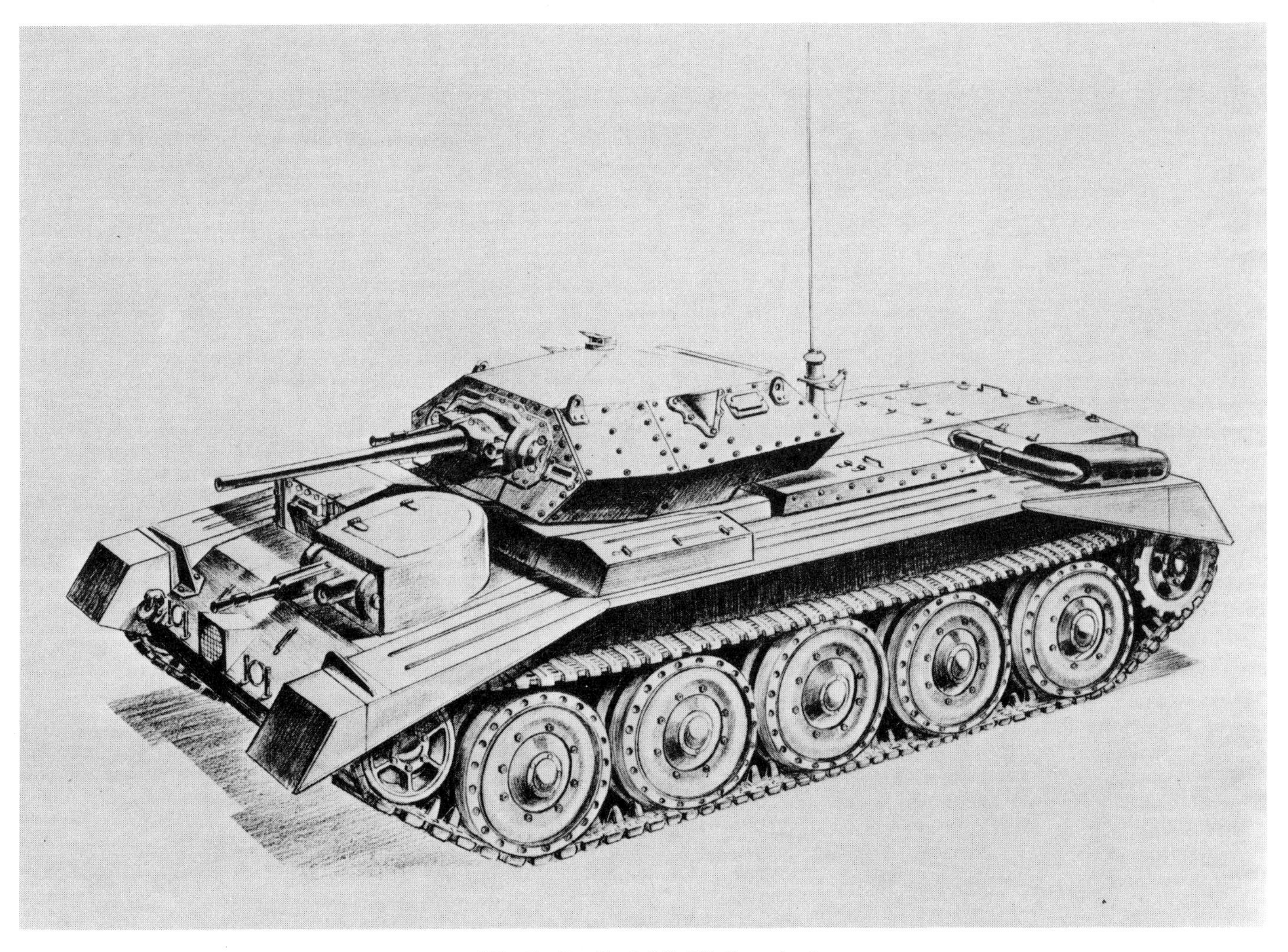

523. Cruiser Tank Mk VI, Crusader I.

APPENDIX 1

British and American Tank Guns, Engines, and Fittings

WHILE it is beyond the scope of this book to give full technical descriptions of tank components and tank engineering, it is desirable to provide additional details of principal features of armament and chassis of British and American tanks of World War II. For the sake of brevity, and to avoid constant repetition, full details of individual weapons, engines, and so on, are omitted from the specifications given with each vehicle. The reader will find more comprehensive data for the most important features here.

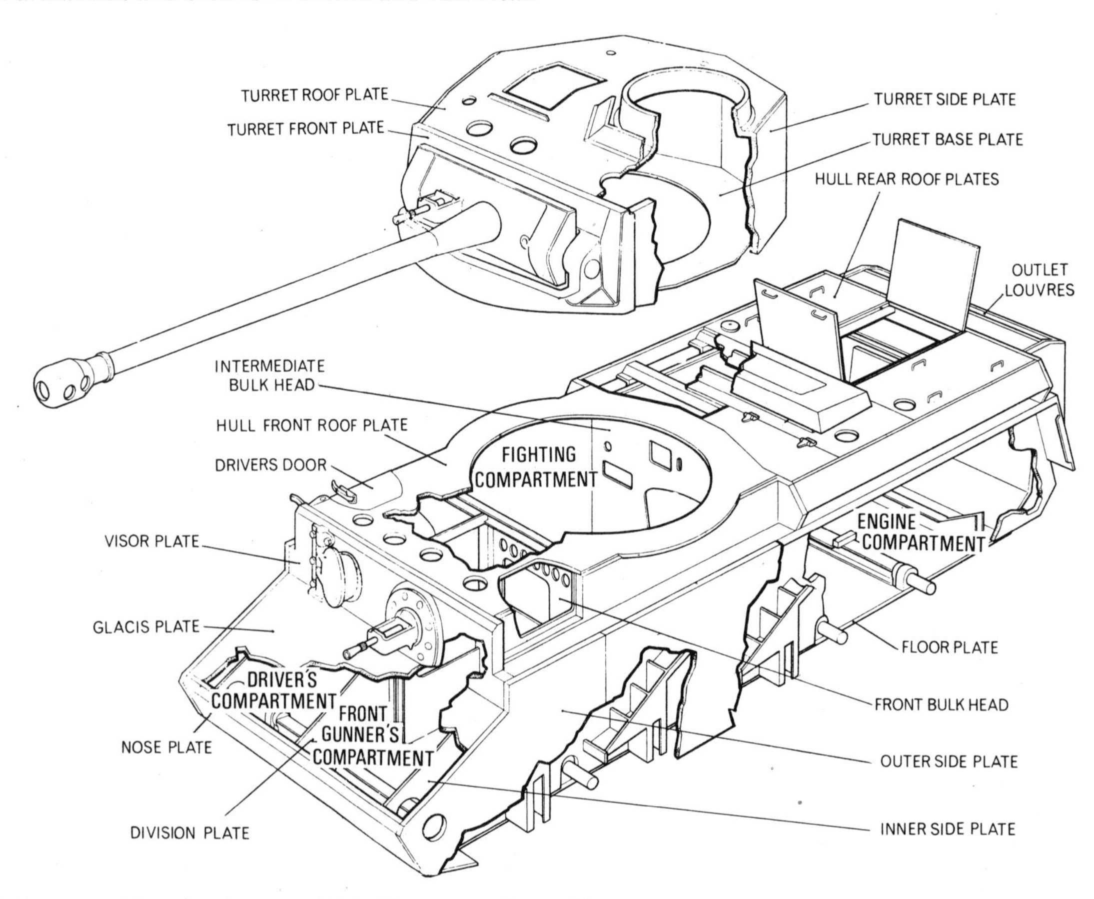

1: General layout and interior features (524: The parts of a tank).

This "exploded" diagram shows the most common terms used to describe the structure of the hull and turret, and such terms are to be found in the text of this book. Not marked in the drawing are the following:

Cupola—raised hatchway, usually for the commander, containing vision devices by which means he conns and directs the vehicle when in fully "closed down" action state. In many instances the cupola rotated independently of the turret. In the drawing, the opening for the cupola is seen (left) on the turret roof.

Mantlet or **mantle**—armoured housing protecting the gun mount in the turret front. This can be seen on the turret in the drawing.

Mount—the fitting holding the gun in position in the turret; in many cases the mount and mantlet are integral and in American terminology "mount" is synonymous with "mantlet".

Muzzle Brake—an arrangement of baffles at the end of the barrel of the gun intended to reduce recoil forces in high velocity guns. This can be seen on the gun in the drawing.

Counter Weight—(1) a weight on the muzzle of a gun (sometimes giving the appearance of a muzzle brake) to balance the weapon; (2) a weight on the rear of a turret to balance the weight of the gun.

Rear Decking, Front Decking—alternative terms for the hull rear and front roof plates shown in the drawing.

Turret Ring—(1) the circular opening in the hull roof in which the turret fits; (2) more precisely, the ball race around the opening on which the turret traverses.

Turret Cage or Basket—lower fixed section of turret which fits inside the hull within the turret ring, rotating with the turret and forming a floor and working space for the turret crew; not always fitted and not shown in the drawing.

Vision Slot—opening in hull or turret providing view outside the vehicle for crewmen when in the fully "closed down" action state; usually provided with protective cover and suitable vision blocks.

Pistol Port—small opening in hull or, more usually, turret side protected by a hinged flap enabling crewmen to either fire with small arms or communicate with personnel outside the vehicle when the tank is in a "closed down" state; pistol ports were also frequently used to discharge empty cartridge cases from inside of the vehicle.

Escape Hatch—hatch in belly, hull side, or other part of vehicle specifically intended for crew exit in an emergency.

Sponson—term for the outer sections of a vehicle's superstructure on each side, usually overhanging the track or suspension, hence the term "sponson-mounted gun"; sponsons were sometimes called panniers.

Superstructure—that part of the hull standing above the level of the tracks or track covers.

Chassis—the lower part of the hull carrying the suspension and drive units.

In addition to the above, the terms "Front Gunner", "Bow Gunner", "Hull Gunner", "Assistant Driver", and "Co-driver" are frequently encountered, and these all generally refer to the same crewman, whose task, in combat at least, was to operate the machine gun fitted in the front hull. The US Army favoured the term "Assistant Driver", but in this book the alternative "Co-driver" has been used for brevity.

525.
Cut-away view of a Churchill VII shows typical interior layout of British tanks. Note in the front compartment the driver's and front gunner's seats, the hull machine gun, the driving controls, and the side escape hatch; turret has gunner's seat and sighting equipment, wireless at the rear, commander's cupola and loader's hatch in roof, and rear stowage box, plus—in this case—a turret cage. In this vehicle the drive is taken to sprockets at the rear and the front idlers are used for tensioning the track, the adjusting screw for this being clearly visible at the extreme left. Also worthy of note are the periscopes in the turret and front hull roofs, the periscopes in the cupola, and the adjacent handles by means of which the commander rotated the cupola.

526.
The more compact interior layout of a typical light tank, in this case a Harry Hopkins, is seen in this view. Note the very low driver's seat, the foam rubber roof lining above it, and the driver's periscope. In the turret can be seen the ammunition stowage, the gunner's seat, his telescope sight, and the wireless set mounted in the rear overhang. This particular type had steering wheel instead of steering levers. The steering mechanism for the road wheel axles (a feature peculiar to the Tetrarch/Harry Hopkins vehicles) and control leads to the gearbox may be seen in the belly of the tank.

APPENDIX 1

Interior of Medium Tank M4A4 (Sherman V) shows similar layout arrangement to that of the Churchill (plate 525) and is characteristic of American tanks. Note that in this instance the drive is to sprockets at the front, necessitating a drive shaft passing forward to the front-mounted transmission. With the exception of the early light tanks, all British tanks of World War II had the drive to rear sprockets, a feature subsequently adopted for the medium tanks developed after the M4 series by the United States. Note that the M4 has the wireless equipment carried in the turret rear, a feature copied from (and suggested by) the British in this design. Previous American types had the wireless installed in the hull. The M24 series light tank also featured rear drive.

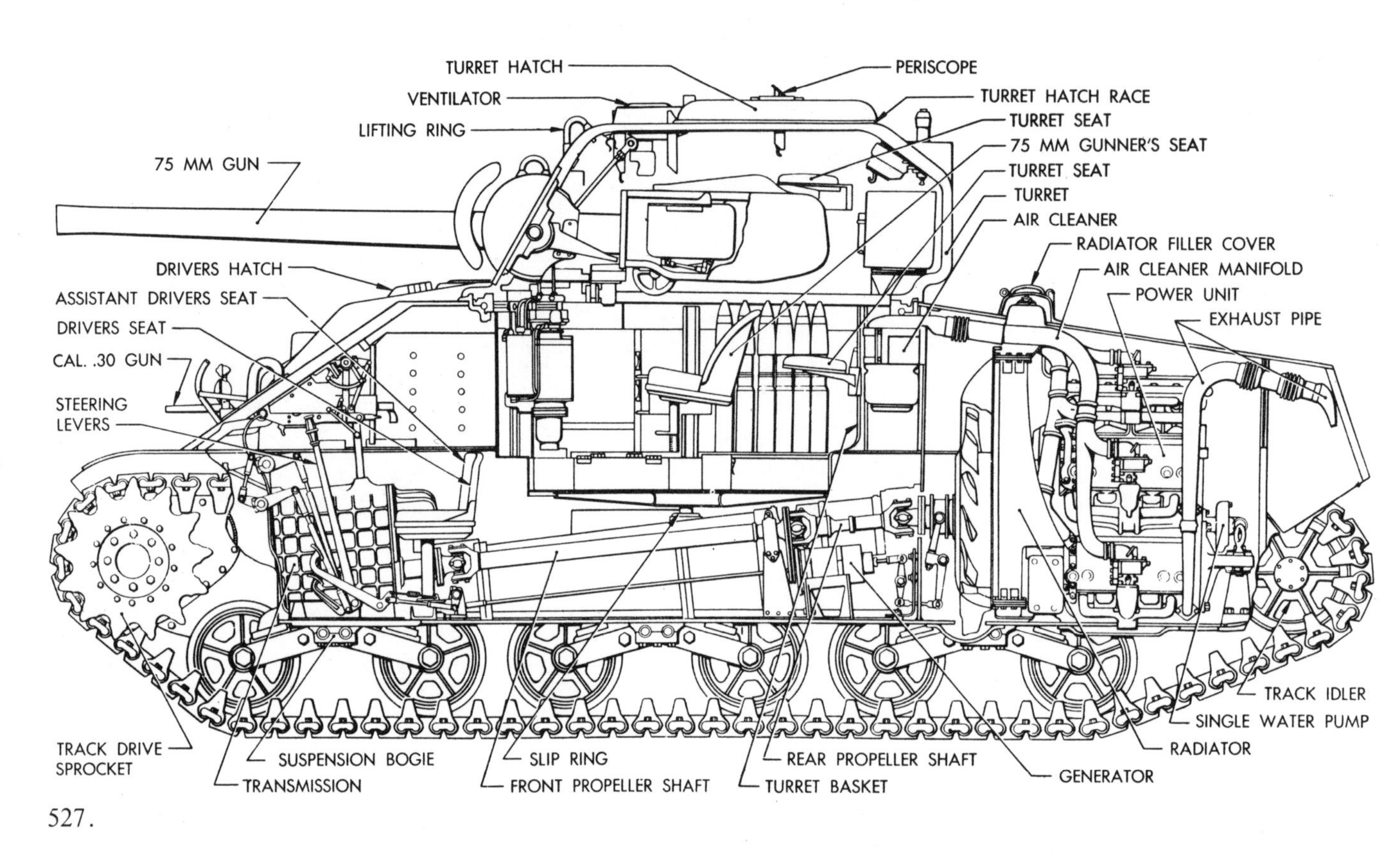

527.

2: Principal Tank Guns

2pdr Ordnance Quick Firing Mks IX, X, and XA
(Illustration: see Plate 526) **United Kingdom**

Adopted for tank and anti-tank use in January 1935, this weapon was the heaviest gun fitted to British tanks from 1939-42, when the 6pdr began to supplant it in new production. While an effective and accurate weapon, and superior to the German 37mm tank gun in penetrating power, the 2pdr proved inadequate by 1941 partly due to the introduction of 50mm, 75mm and 88mm guns in the tank and anti-tank roles by the Germans, all of which could outrange it, partly due to superior German tactics as a result, and partly due to its inability to fire HE (high explosive) shells. Sometimes, also, 2pdr shot broke up on German face-hardened armour. To improve the 2pdr's performance an APCBC (armour piercing capped, ballistic capped) shot was introduced in September 1942. In 1940 a Czech designer, Janecec, had suggested a reduced bore attachment for the 2pdr designed to increase the muzzle velocity, and thus the range and penetrating power. No urgency was given to adopting this, however, until it was discovered that the Germans were working on a similar device. By the time the Janecec idea was approved—as the Littlejohn adaptor—in January 1943, the 6pdr gun was coming into service in tanks and the 2pdr was rapidly being replaced.

Weapon details:
Calibre: 1·575in (40mm)
Length of tube: 78¾in
Overall length: 81·95in
Weight: 287lb
Muzzle velocity: 2,800fps (AP), 2,600fps (APCBC)
Weight of shot: 2lb 6oz, 2lb 1½oz (APCBC)
Armour penetration: 57 (AP) and 57·5mm (APCBC) at 500 yards/30°
Vehicles fitted: A9, A10, A13, Covenanter, Crusader I & II, Matilda, Valentine I-VII, Churchill I & II, Ram I, Sentinel AC I, Tetrarch, Harry Hopkins.

6pdr Ordnance Quick Firing Mks 3 and 5
United Kingdom

Design of the 6pdr as both a tank and anti-tank gun to replace the 2pdr was initiated by the Director of Artillery in April 1938. However, low priority was given to development due to pre-occupation with other design work, and a prototype was not ready until Spring 1940. Initial orders were placed in December 1940 and mass-production was ordered in February-May 1941. By this time the need for a more powerful tank/anti-tank gun had been clearly demonstrated by events in the Western Desert fighting. The first 300 6pdr's were delivered in December 1941, though no tanks with this weapon were ready until March 1942, when the Churchill III was produced, followed by the Crusader III in May 1942.

There was considerable delay in getting tanks with 6pdr guns into production, initial efforts being concentrated on a new vehicle (the A27) to take this gun, followed by a decision to adapt the existing Churchill and Crusader when it was appreciated that the A27 would not be ready for many months. In each case no work on these proposals was started until *after* the 6pdr had been ordered into production with the result, as seen above, that the guns were ready long before there were any tanks to mount them.

Though the 6pdr gave increased hitting power to British tanks, it still suffered from the tactical disadvantage of not being able to fire a HE round which meant that British tanks were still largely dependent on field artillery for fire support, an unsatisfactory state of affairs, especially in the open fighting of the Western Desert. Attempts to give the 6pdr—which was a good accurate weapon—improved penetrating power led to the development of Discarding Sabot ammunition for it in early 1944.

The 6pdr Mk 5 was distinguished from the Mk 3 by its noticeably longer barrel of "lighter" appearance. It was usually fitted with a counterweight on the muzzle.

Weapon details:
Calibre: 2·24in (57mm)
Length of tube: 96·2in (Mk 3), 112·2in (Mk 5)
Overall length: 100·95in (Mk 3), 116·95in (Mk 5)
Weight: 768lb (Mk 3), 720lb (Mk 5)
Muzzle velocity: 2,800fps (Mk 3), 2,965fps (Mk 5)
Weight of shot: 6lb 4oz Armour penetration: 81 (Mk 3) and 83mm (Mk 5) at 500 yards/30°
Vehicles fitted: Crusader III, Churchill IV, Churchill III, Cavalier, Centaur I, Valentine VIII, Cromwell I and II, Ram II.

75mm Ordnance Quick Firing Mks V and VA
United Kingdom

(Illustration: see plate 525, gun mounted in Churchill VII)
Appearance of the American M3 and M4 medium tanks in British service in late 1942 showed up the inadequacies of the 6pdr. The 75mm gun mounted in these American-built vehicles was the first tank gun used by the British which had a "dual purpose" capability—it could fire HE and AP shot as required, thus matching German tactical flexibility in this respect. As a result of this, the British General Staff asked, in December 1942, if similar 75mm guns could be mounted in British tanks and the Ministry of Supply designed a suitable weapon, based on the American gun, which would fit the existing 6pdr mount and fire American ammunition. First vehicle mounting the 75mm gun, the Cromwell IV, appeared in October 1943 and plans were also being made to mount the weapon in the Churchill. At first, however, the 75mm gun proved unsatisfactory due mainly to an inadequate mounting, and teething troubles with this weapon were not finally overcome until March 1944. For the invasion of Europe in June 1944, the 75mm MV (medium velocity) gun was the principal British tank gun for the Cromwell and Churchill, but by then had been outmoded by better German tank guns, and German tanks (eg, the Panther) which had superior armour thickness.

Weapon details:
Calibre: 2·953in (75mm)
Length of tube: 107·8in
Overall length: 112·576in
Weight: 692lb
Muzzle velocity: 2,030fps
Weight of shot: 13¾lb (HE and AP)
Armour penetration: 68mm at 500 yards/30°
Vehicles fitted: Centaur III, Cromwell IV-VII, Churchill VI and VII, Valentine X & XI, A33.

528. 6pdr gun, prototype shown in test mount.

17pdr Ordnance Quick Firing Mks II, IV, VI and VII
(see plate 529) **United Kingdom**

Like the 2pdr and 6pdr before it, the 17pdr was designed from the start as both a tank and anti-tank gun. It was first proposed in early 1941 and designs were approved in Summer 1942. Pending availability of the US M10 GMC with its 3in gun as a tank destroyer vehicle for delivery to British forces, it was considered essential to develop a British vehicle with the 17pdr gun. This gave rise to the Challenger (qv) which emphasised the limitations of existing British tank designs, all of which were too narrow to take a turret capable of mounting the 17pdr. A much modified derivation of the Cromwell, the Challenger ultimately proved unsatisfactory and the Churchill also had to be re-designed (as the Black Prince) to take a 17pdr and appeared too late to see service. As a "stop gap" existing M4 Shermans in British service were modified to take the 17pdr, which was just possible by adding a turret extension, turning the gun on its side and adapting it for left hand loading, and re-arranging the internal stowage of the vehicle. Known as the Firefly in this guise, the modified Sherman with this gun was the only British vehicle in service (though in limited numbers) at the time of the invasion of Europe in mid 1944 mounting a gun which was a theoretical match for the German Panthers, Tigers, and late-model PzKw IVs. Subsequently the American-built M10 in British service was also rearmed with the 17pdr (becoming the Achilles), two British-built SP vehicles carried it, and the new A41 (Centurion) was

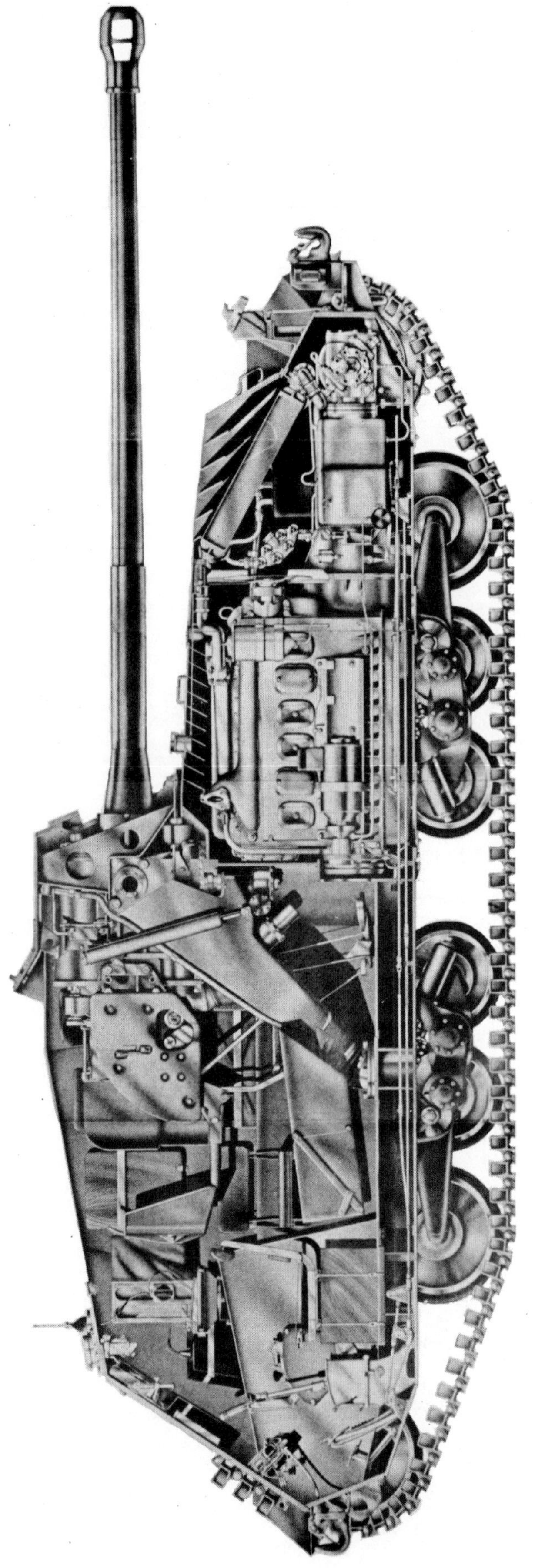

529. 17pdr gun mounted in the Archer SP.

designed around it. For further details, see development histories of the vehicles mentioned above. Discarding Sabot and HE shell for this weapon were developed in 1944.
Weapon details:
Calibre: 3in (76·2mm)
Length of tube: 165·45in
Overall length: 172·25 (Mk II), 184·03 (Mk IV), 183·8 (Mk VI), 184·05 (Mk VII) inches
Weight: 1,885–2,032lb according to mark.
Muzzle velocity: 2,900fps; 3,950 fps. (Mk VII)
Weight of shot: 17lb (AP, APC, APCBC, HE); 8·15lb (APDS)
Armour penetration: 120mm at 500 yards/30° (APDS: 186mm)
Vehicles fitted: Sherman Firefly, Challenger, Black Prince, TOG2*, Achilles (M10), Archer, Centurion I (Mk VII gun only), Sentinel AC IV.

77mm Ordnance Quick Firing Mk II United Kingdom

Developed by Vickers (and known at first as the Vickers HV 75mm gun), this weapon was essentially the 17pdr tailored to fit a turret of Cromwell dimensions. This was mounted in the "improved" Cromwell tank, the A34 Comet. Characteristics were similar to those of the 17pdr, but the gun was shorter and lighter. Its performance was only slightly inferior to that of the 17pdr. Calibre and shot were the same as for the 17pdr.
Weapon details:
Length of tube: 147·65in
Overall length: 165·5in
Weight: 1,502lb
Muzzle velocity: 2,600fps
Armour penetration: 109mm at 500 yards/30°
Vehicle fitted: Comet (other details as 17pdr).

3in Howitzer Ordnance Quick Firing Mk I and 1A United Kingdom

This weapon was fitted to the close support (CS) variants of tanks armed with 2pdr guns. CS tanks were issued on a scale of about two vehicles per squadron (company) of 15. Their main function was to fire smoke shells but they could also fire HE.
Calibre: 3in (76·2mm)
Length of tube: 75in
Overall length: 78·2in
Weight: 256lb
Muzzle velocity: 600–700fps
Weight of shot: 13¼/13⅞lb (Smoke and HE)
Range: 2,000–2,500 yards
Vehicles fitted: Tetrarch ICS, Churchill I, Matilda CS, Valentine III (New Zealand), Covenanter CS, Crusader CS.

3·7in Mortar (Howitzer) Ordnance Quick Firing Mk I United Kingdom

This was another howitzer type gun, fitted to some CS models of 2pdr armed British tanks. It fired Smoke only.
Calibre: 3·7in (95mm)
Length of tube: 55½in
Overall length: 59·2in
Weight: 222lb
Muzzle velocity: 620fps
Weight of shot: 10½lb
Vehicles fitted: CS versions of A9, A10, A13.

95mm Tank Howitzer Ordnance Quick Firing Mk I — **United Kingdom**

Final type of howitzer used in British CS tanks was the same calibre as the 3·7in mortar, but much more refined, firing Smoke, HE, and HEAT (high explosive, anti-tank). It had a counterweight on the muzzle to balance the 6pdr type mounting in which it was fitted. This weapon armed the CS versions of tanks with 6pdr guns and was last of its line, since tanks with 75mm guns could fire their own HE.
Calibre: 3·7in (95mm)
Length of tube: 80·4in
Overall length: 85·52in
Weight: 867lb
Muzzle velocity: 1,075fps
Weight of shot: 25lb
Armour penetration: 110mm at 500 yards/30° (HEAT only)
Vehicles fitted: Centaur IV, Cromwell VI & VIII, Churchill V & VIII, Alecto I.

530. 95mm Tank Howitzer in Churchill VIII.

Mortar, Recoiling Spigot, Mks I and II — **United Kingdom**

Work on a special demolition gun was started in September 1942, when lessons of the Dieppe raid had been evaluated and indicated the need for short range bomb throwing weapon for the demolition of pillboxes and anti-invasion obstacles. By 1943, the Churchill had been chosen for conversion to the AVRE role and the Petard mortar was adapted to fit the 6pdr gun mount. Churchills armed with this weapon took part in Operation Overlord, the D-Day landing and proved highly effective in the assault engineer role. The weapon was "broken" for loading, outside the turret, a special sliding loading hatch being provided in the front hull roof. The demolition bomb was popularly known as the "Flying Dustbin".
Calibre: 290mm
Length of tube (loading trough): 45in
Overall length: 86in
Weight of bomb: 40lb
Effective range (with demolition bomb): 80 yards
Vehicles fitted: Churchill III & IV (Mk I mortar used in training vehicles only).

531. The Petard (Recoiling Spigot) Mortar fitted to the Churchill AVRE.

37mm Gun M6 — **United States**

The 37mm tank and anti-tank gun was developed early in 1938, earmarked initially for installation as main armament for the M2 series medium tanks. Production of the tank version of this weapon started at Watervliet Arsenal, NY, under the designation M5. M6 was an improved version introduced in 1941 and earlier designation numbers were allocated to the anti-tank version. In mid 1940 additional contracts were placed with commercial plants to meet the greatly increased requirement for 37mm guns for fitting to M2A4 and M3 series light tanks. From 1941 onwards the 37mm gun was used in conjunction with a gyrostabiliser in both light and medium tanks. It was a good serviceable weapon, but obsolescent from 1940 in the face of the adoption of 50mm and 75mm guns by the Germans.
Weapon details:
Calibre: 1·457in (37mm)
Length of tube: 72·85in
Overall length: 77·35in
Weight: 185lb
Muzzle velocity: 2,545fps
Armour penetration: 48mm at 500 yards/30°
Vehicles fitted: Medium Tanks M2, M2A1, M3; Light Tanks M2A4, M3 series, M5 series, M22.

532. 37mm Gun M6 on an M5 Light Tank.

APPENDIX 1

75mm Gun M2 and M3 **United States**

By May-June 1940 it had become clear that it would be essential to mount a 75mm gun in new American medium tanks in order to match the heavier calibre guns being produced by the Germans for their tanks. The M3 medium tank was designed round a 75mm gun, the layout being based on the experimental T5E3 of 1939 which had mounted a 75mm howitzer in the hull. Watervliet Arsenal quickly designed a 75mm gun and mount, based broadly on the French 75mm gun which had been standard US Army field equipment since 1918. In mid July 1940 a first order for 1,308 guns and mounts was given to Watervliet and the first of these was delivered in April 1941 just in time for fitting to the early M3 medium tanks. Subsequent demand for 75mm guns became so great that two commercial firms were given contracts for producing the 75mm gun in late 1940, followed by a third late in 1941. Ultimately more than 20,000 M2 and M3 75mm guns were produced by 1943-44. The American 75mm gun could fire both AP and HE shells. It was sturdy and simple to operate with a performance which, for the first time, gave the British superiority in tank gunpower over the Germans when the American medium tanks first appeared in British hands in the Western Desert in 1942. In the M3 medium series, the mount used with the gun was designated M1, and in the M4 series, the mount was designated M34, and, in improved form, M34A1. The M34A1 mount was distinguished from the M34 mount by its full width armour shield with "ears" flanking the barrel. The M2 gun was shorter than the M3 version and was frequently seen with a prominent counterweight on the muzzle, necessary to balance the gyrostabiliser by simulating the weight of the longer M3 gun.

Weapon details:
Length of tube: 84in (M2), 110·625in (M3)
Calibre: 75mm
Overall length: 91$\frac{3}{4}$in (M2), 118·375in (M3)
Weight: 783lb (M2), 910lb (M3)
Muzzle velocity: 1,860fps (M2), 2,300fps (M3)
Armour penetration: 60mm (M2) and 70mm (M3) at 500 yards/30°
Vehicles fitted: Medium Tanks M3 series (M2), later vehicle (M3), Medium Tanks M4 series (M3), very early vehicles (M2), Churchill IV (NA75) (M3).

533. 75mm gun M3 and M34A1 mount from M4A4 Sherman.

Gyrostabiliser

Revolutionary feature of American tank gun control introduced with the M3 light and medium tanks was the gyrostabiliser which allowed the gun to maintain its aim (in elevation) while the vehicle was moving over rough terrain. Previous to its introduction a tank gun could only be aimed with accuracy while the tank was stationary, thus presenting a possible target to the enemy. The gyrostabiliser was based on the gyroscope principle whereby the gyro was attached to the gun cradle on the same axis as the gun and set spinning. Displacement of the gyro by vertical movement of the tank caused the stabiliser to return the gun barrel to its original axis, thus maintaining any given elevation irrespective of the up-and-down movement of the vehicle. The idea was taken from a similar system used in naval gunnery. For tanks, however, several problems had to be overcome, particularly that of reducing the system to fit inside a tank, tough enough to withstand the rigours of combat. Neither British nor German tanks had this feature, and American tanks from the M3 light and medium on had a definite tactical advantage in this respect, though gyrostabilisers were not, in fact, used as frequently as they could have been in combat. The picture shows a gyrostabiliser installed in a medium tank complete with its control box and attached to the arm of the cradle.

534. Gyrostabiliser installation in a M4 Medium Tank turret.

535. 90mm gun T7 (prototype for the 90mm gun M3) shown mounted in the T71 GMC which was pilot model for the M36 GMC.

76mm Gun, M1, M1A1, M1A1C, amd M1A2 United States

American observation of the fighting between the British and Germans in the Western Desert in 1942, indicated that a gun even more powerful than the 75mm M3 weapon was desirable. In August 1942, the US Ordnance Department designed a new high velocity gun able to fire the existing 3in anti-tank shell, keeping overall weight of the piece as low as possible. This new gun was produced as a prototype in less than a month, being test fired in an M4 medium tank in September 1942. It was standardised instantly as the 76mm gun M1 and put into limited production. As is indicated elsewhere in this book, there was some difficulty in getting the using arms to accept the heavier gun and it was not installed in M4 tanks until 1944, after which Army Ground Forces clamoured for more. The 76mm gun was roughly comparable to the British 17pdr, but had inferior penetrating power. However, it had nearly twice the rate of fire and was simpler to operate and maintain. The various models represented successive improvements, the M1A1C and M1A2 having muzzle brakes. All were to be seen on M4 series vehicles from 1944 on. The mount was designated M62.

Weapon details:
Calibre: 76mm (3in)
Length of tube: 156in
Overall length: 168in
Weight: 127lb
Muzzle velocity: 2,600fps (APC), 3,400fps (HVAP), 2,700fps (HE)
Armour penetration: 88mm at 1000 yards/30° (APC), 133mm at 1000 yards/30° (HVAP)
Vehicles fitted: Medium Tanks M4 series (late production), T23, M18 Hellcat.

90mm Gun M3 United States

An adaptation of the 90mm AA gun, development of which had started early in 1942, the prototype of the 90mm tank gun was ready by December 1942. It was test-fired in the turret of an M10 GMC early in 1943 and this was subsequently to lead to development of the M36 GMC as a tank destroyer specifically to mount the weapon as a replacement for the M10. There was some opposition to the acceptance of a gun of this calibre in the using arms, as indicated elsewhere in this book. In 1944 the 90mm gun was fitted to the T25/T26 series medium/heavy tanks, and the M26 tank with the M3 gun eventually went into service early in 1945. The M36 series with the 90mm gun entered service in mid 1944. The 90mm M3 gun had an inferior penetrating power to the British 17pdr and was less accurate. However it was simpler to install, operate, and maintain. The 90mm gun was similarly inferior to the German 88mm gun, though the most powerful tank gun in service in the US Army in 1945.

Weapon details:
Calibre: 90mm
Length of tube: 177·25in
Overall length: 195in
Weight: 2410lb
Muzzle velocity: 2,800fps (APC), 3,350fps (HVAP), 2,700fps (HE)
Armour penetration: 120mm at 1000 yards/30° (APC), 195mm at 1000 yards/30° (HVAP).
Vehicles fitted: M36 series, T25/T26, M26 Pershing.

APPENDIX 1

3: Tank Machine Guns

Most British and American tanks carried a secondary armament in the form of machine guns, to supplement the main armament and to provide local AA protection. Typical disposition was a single flexibly-mounted weapon in the hull front to engage, for instance, infantry, a single machine gun fixed in the mantlet co-axial with the main armament and elevating and traversing with it, plus a flexibly-mounted machine gun on the turret roof or cupola, sometimes in a ring-mount, for AA defence. This latter was not always mounted, and some tanks did not have all three, while others had more. Most important types of machine guns carried in British and US tanks are detailed below:

UNITED STATES

Browning ·50 cal machine gun M2, heavy barrel

This weapon was the normal AA machine gun for medium and heavy tanks, and the larger gun motor carriages. It was flexibly mounted on either a fixed pintle or a ring-mount.
Calibre: ·50 in
Feed: Disintegrating link belt
Length: 5ft 5in
Rate of fire: 450–550rpm

Browning ·30 cal machine gun M1919A

This weapon was the usual hull front weapon in a flexible mount and formed the AA machine gun in light tanks. A fixed version was the normal co-axial machine gun in American tanks. This ·30 cal weapon was also used in some British tanks.
Calibre: ·30in
Feed: Disintegrating link belt
Length: 3ft 6in
Rate of fire: 450–600rpm

BRITISH

Bren ·303 cal light machine gun

This weapon was used singly or in pairs on a special mount on the turret of some British tanks for local AA defence.
Calibre: ·3·3in

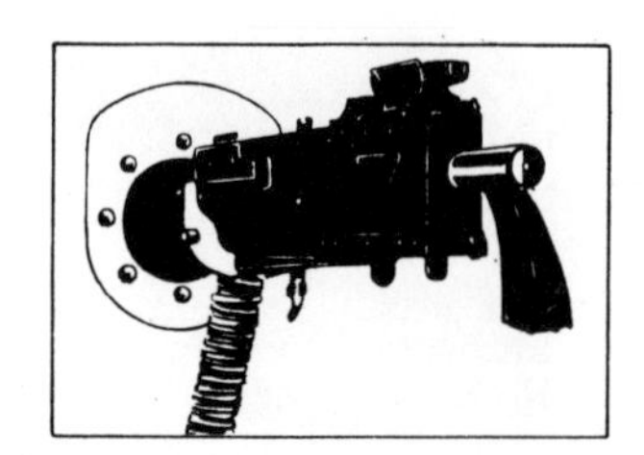

536. Typical mounting for M1919A Browning ·30 cal machine gun. For typical turret mounting of Browning ·50 cal machine gun see plate 535.

Feed: Magazine or drum
Length: 3ft 9in
Rate of fire: 450–550 rpm

Besa 7·92 cal tank machine gun Mk I, II, and III

This was the principal machine gun type used in British tanks.
Calibre: 7·92mm
Feed: Belt, 225 rounds
Length: 3ft 7½in
Rate of fire: 750–850 rpm

Vickers ·303 cal tank machine gun Mk VI, VI*, VII

This weapon was used in the earlier British tanks in the hull or co-axial positions but was largely superseded by the Besa described above.
Calibre: ·303 in
Feed: Belt
Length: 3ft 8in
Rate of fire: 500 rpm

Vickers ·50 cal tank machine gun Mk V

This weapon was mainly used in light tanks.
Calibre: ·50 inch
Length: 3ft 11½in
Rate of fire: 350–400 rpm

4: Smoke Devices

In order to provide local smoke cover to conceal a vehicle in an emergency, all British tanks from 1939 on were provided with single or twin 4in smoke dischargers on the turret sides. These were also fitted to American-built vehicles in British service. The dischargers held smoke candles which were fired electrically by the vehicle commander. They could only be reloaded from outside the vehicle, however. Some tanks had a similar device fitted at the rear of the hull, when it was known as a "rear smoke emitter". The Valentine, Churchill, A27 series, and all later British tanks were fitted with a 2in bomb thrower in the turret roof, a small mortar specifically to fire smoke bombs for which up to twenty "ready use" rounds were carried, thus allowing a longer smoke screen to be laid. The Canadian Ram and Grizzly tanks also had bomb throwers of this type, and a similar weapon was subsequently fitted in the turret roof of M4 medium tanks, and all later US types, from the start of 1944. This was a direct result of the example set in the Grizzly.

537. 4in Smoke Discharger on turret of Light Tank Mk VIC.

5: Principal Tank Engines

Mainly for economy reasons most British pre-war tank designs utilised adapted commercial engines; an exception were some of the early Vickers tanks, some of which were powered by a purpose-built engine, developed by an associate company, Armstrong-Siddeley, which came with either V8 or V12 configuration. Main limitation of commercial engines, however, was their relatively low power output which restricted vehicle size. With larger, heavier, vehicles more power was necessary and this was achieved in the cruiser tanks A13-A27L by using the American Liberty engine, modified by Nuffields the British licensees. This was basically a 1918 vintage aero engine (as used by Christie in his high speed tanks) which gave an excellent power output but was bulky and technically outmoded. This was replaced in later cruiser tanks by the Rolls-Royce Meteor which was similarly adapted from an aero engine, the Rolls-Royce Merlin which had powered the Spitfire fighter. A most excellent power plant with a good reserve of output, it became the standard British tank engine in the latter half of the war. It was installed in most new designs from the A27M (Cromwell) onwards. The other expedient for obtaining more power in British tanks was to use two or more commercial engines "ganged" together. Examples were the Bedford Flat-12 (Churchill) and AEC twin diesels used in some marks of Matilda. The Meadows engine used in the Covenanter also fell in this category. With a few exceptions, all British tank engine were gasoline (petrol) types.

In the United States, prewar fiscal limitations had also restricted the choice (and development) of power plants for tanks. The Wright Continental W-670 radial air-cooled engine, another aero engine with good power output, was the most favoured type and was used in most American light tanks of prewar construction. With the coming of heavier vehicles, however, like the M2 and M3 mediums, more power was necessary and the larger Wright Whirlwind radial aero engine was adapted for tank installation as the Continental R-975. Owing to the haste with which it was adopted, however, there were several shortcomings experienced with this engine, among them limited accessibility for maintenance, and cooling problems which caused excessive oil consumption. However by 1941 the demand for tanks was so urgent that there was no ready alternative developed until circumstances forced a change. With urgent need of Wright engines in the aero industry itself, it became necessary to find alternative power plants from within the US automotive industry. This led to the development of several commercial types, some "ganged" together to provide sufficient power. Among them was the General Motors twin diesel two-stroke engine, the Chrysler A-57 multi-bank five-line engine (five crankshafts with the engine outputs geared to a common output), the Guiberson diesel radial, and the Ford GAA V12/V8 petrol engine which was, in fact, developed from a design for an aero engine. Of these the Guiberson was used as an alternative to the Continental W-670 in light tanks and was the first to be discarded when the supply situation eased slightly in 1942. The GM diesel unit was not liked by the US Army, but was much favoured by users in the field, despite several mechanical limitations which necessitated careful maintenance. Much of the fond regard for the diesel powered tanks was possibly psychological since legend had it that these vehicles were less prone to catch fire when hit than petrol engined tanks. (In fact most fires were started by hits in the ammunition stowage). Nonetheless the GM diesels were good, giving good torque at low speed and being most reliable when

538. Liberty Mk V engine installed in a Crusader tank. This was a Nuffield-developed version of the original American Liberty engine. It was a water-cooled V-12 unit with a typical output of 395HP at 1,650rpm.

539. Best British tank engine was the Rolls-Royce Meteor which was basically a Merlin aero engine with the supercharger removed, minor modifications for tank use, and a reduced rating. Typical output was 600HP at 2,550rpm.

540. Wright Continental R-975 engine was a radial air-cooled unit, based on the Whirlwind aero engine and was the main American tank engine for the M3 and M4 mediums in 1941.

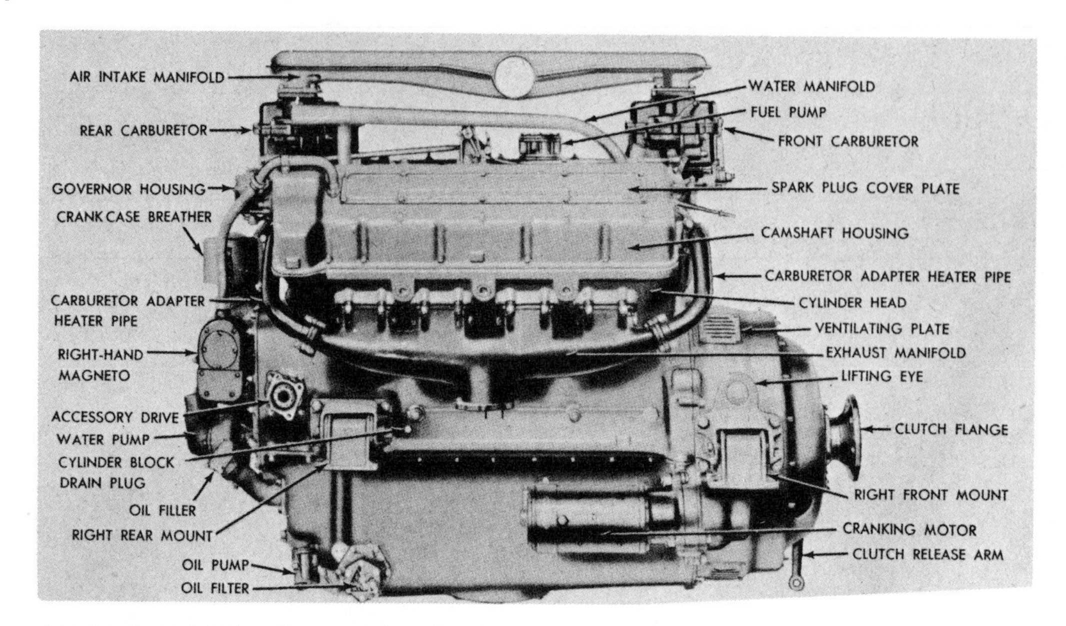

541. Ford GAA V-8 engine was ultimately selected as the "standard" power unit for US medium and heavy tanks. Typical output was 450HP at 2,600rpm compared to 353HP at 2,400rpm with the Continental R975 shown in plate 540. A modified version of the GAA, the GAF, powered the M26 tank.

properly maintained. Most diesel-powered medium tanks (M4A2) were allocated to Lend-Lease stocks, however, by the United States. The Chrysler A-57 engine, developed in great haste in a mere four months also proved to be very reliable, economical and powerful, but it was discarded at an early opportunity, due mainly to maintenance problems posed by the quintuple arrangement of the engines. The Chrysler unit powered the M4A4 and one version of the M3 medium. A Caterpillar engine for installation in M4 series tanks (fitted in the M4A6) was also very promising, but was a late-comer to the scene and was discarded once standardisation had been chosen for the Ford GAA engine. The Ford engine proved to have good accessibility and low fuel consumption as well as high power output and torque. While not perfect, it was considered the best available, and the most adaptable. From late 1943 it became the "standard" American tank engine, though supply never met demand sufficiently for it to displace other types. Variants were produced for tank designs subsequent to the M4 series, but the best-known model was the GAA as fitted in the M4A3 medium tank.

See also:

Plate 525: Bedford Flat-12 engine in Churchill VII tank. This was made up of two 6 cylinder truck engines coupled together. Typical output: 350HP at 2,200rpm.

Plate 526: Meadows Flat-12 engine in Harry Hopkins light tank, an example of another adaptation of a commercial type, designated Meadows M.A.T./3. Typical output: 148HP at 2,200rpm. Same engine was fitted in the Tetrarch tank, and a larger version in the Covenanter.

Plate 527: Chrysler A-57 Multibank five-line engine in medium tank M4A4, a prime example of a power plant in American tanks adapted from commercial sources. It consisted of five Chrysler automobile engines mounted on a common crankshaft. Typical output: 370HP at 2,400rpm.

Plate 529: Twin General Motors 6-71 diesels installed in Archer SP 17pdr.

6: Typical Suspension Systems

"Slow Motion" Type

Designed by Vickers and incorporated in their A9, A10, and Valentine designs, each bogie consisted of three wheels on a pivoted beam with a coil spring.

Horstmann Type

Another Vickers design, used on their light tanks, each bogie consisted of wheels on bell cranks, sprung against each other horizontally. In this form the suspension was restricted to light tanks, but Horstmann suspension was once again used on the Centurion (A41) using the same principal for springing the bogies.

Christie Type

Patented by J. Walter Christie and used on his inter-war fast tanks, this system allowed for independent suspension

542. Close-up of Valentine bogie showing 'slow motion" suspension arm.

543. Horstmann type suspension on Vickers Mk VI Light Tank.

of each road (or bogie) wheel. Each wheel was attached to a pivoted bell crank with a heavy coil spring "damper" anchored to the hull interior and protected by suitable casing. It was used on all British cruiser tanks developed from the A13 which in turn had been based on the Christie tank. The major drawback of Christie suspension was that it took up space inside the hull, a reason why it was dropped by the US Army. It did, however, allow very fast running.

Vertical Volute Spring Suspension

Shown in this instance on a Medium Tank M4 (105mm), this was the standard US suspension system adopted for light and medium tanks. The two wheels in each bogie were pivoted on arms against a vertical spring which was protected by the bogie carrying bracket. A return roller was mounted either on top of, or behind, the bracket, the entire bogie unit forming a self-contained fitting. Advantages of the vertical volute suspension system included simplicity of production and maintenance and easy replacement of damaged units simply by unbolting the complete assembly from the hull.

Horizontal Volute Spring Suspension (HVSS)

Shown in a test installation on a M4A3 tank, HVSS was designed to replace the vertical volute system. It featured four bogie wheels in pairs (instead of two) with each pair sprung horizontally against each other on the bogie unit. Shock absorbers were also incorporated in each bogie. The object was to give a smoother ride than that offered by vertical volute suspension, and allow the use of wider tracks. HVSS was so designed as to allow any individual wheel to be replaced without disturbing other wheels on the bogie. The return rollers were now set on the hull side as separate fittings. All late M4 series vehicles had HVSS.

Torsion Bar Type

Superficially this looked like Christie suspension, and each wheel was carried on an individual arm as in Christie suspension, but the spring, in the form of a torsion bar was usually carried across the hull and anchored on the far side.

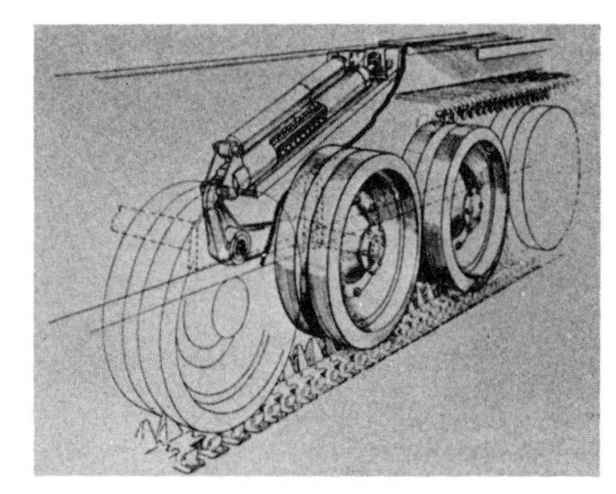

544. Cut-away of Christie suspension system in Cromwell tank.

545. Vertical volute spring suspension on M4 Medium Tank.

546. Horizontal volute spring suspension on M4A3 Medium Tank.

The torsion bars to the wheels on each side were placed alternately across the hull floor. Advantages of this system included better space utilisation, better protection for the springs and relative simplicity. More careful production control of the spring was however needed. Most later types of US medium/heavy tanks (eg, M26) had this suspension.

APPENDIX 2

Comparative diagrams of major British and American tanks and motor carriages

Drawn by Kenneth M. Jones. Scale—1:76 (constant).

The silhouette of a 6ft man is included to give a clear indication of relative sizes. Vehicles are shown in chronological order by type.

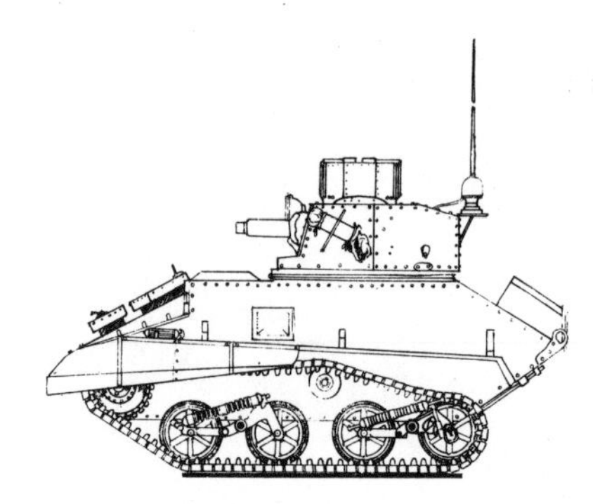

547. Light Tank Mk VIA—Vickers machine guns. 1937.

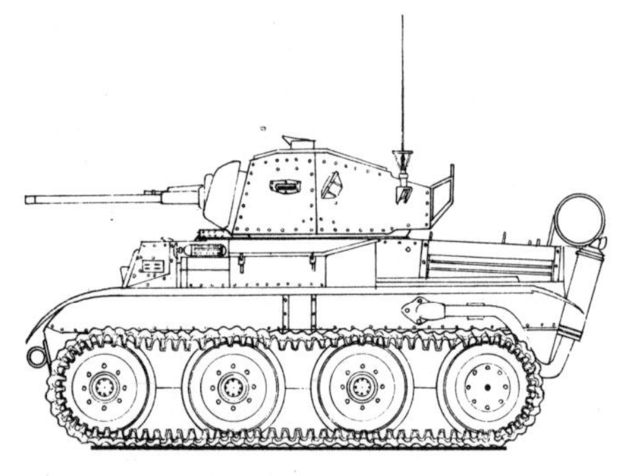

548. Light tank Mk VII, Tetrarch—2pdr gun. 1938–40.

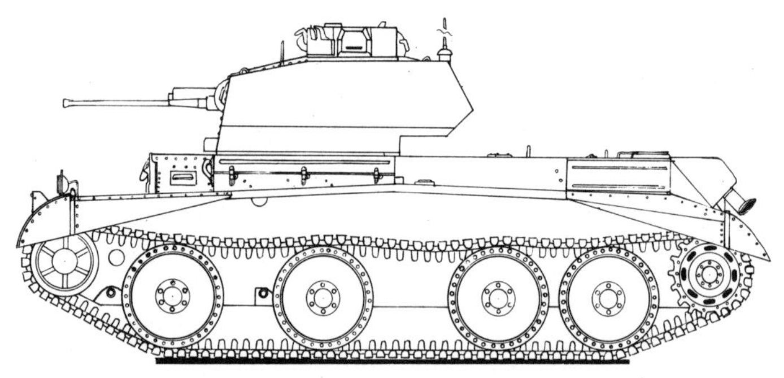

549. Cruiser Tank Mk IV (A13 Mk II)—2pdr gun. 1939.

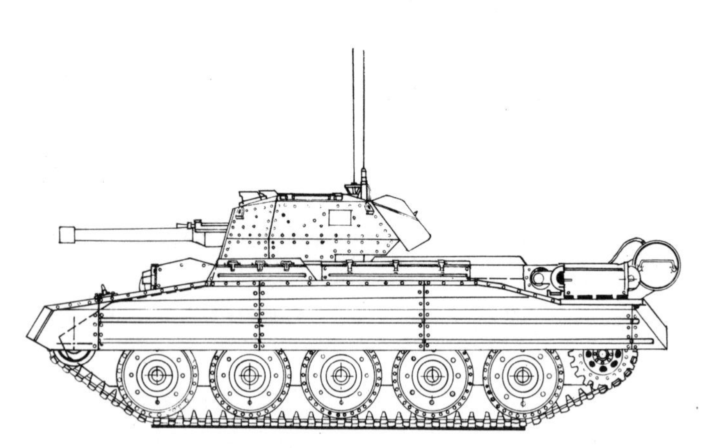

550. Cruiser Tank Mk VI (A15), Crusader III—6pdr gun. 1942.

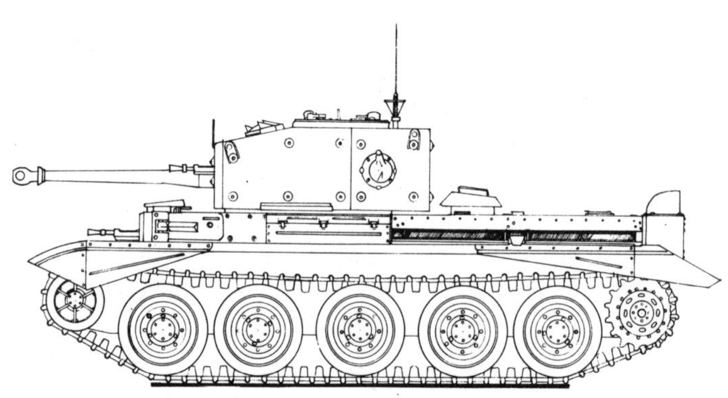

551. Cruiser Tank, Cromwell IV—75mm gun. 1943.

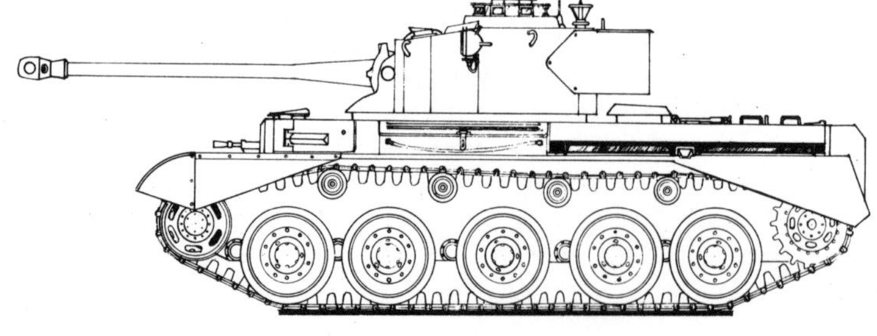

552. Cruiser Tank, Comet—77mm gun. 1945.

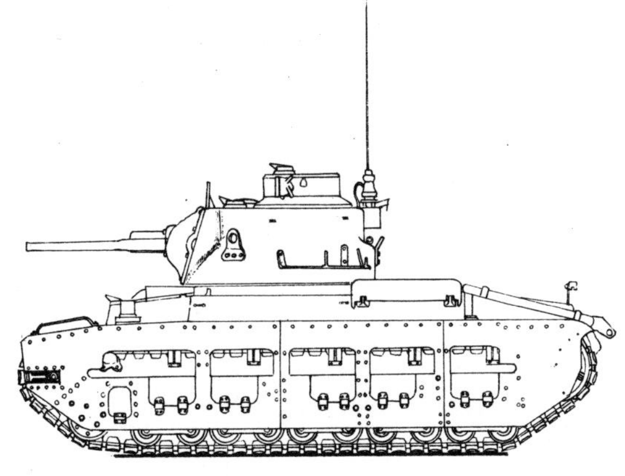

553. Infantry Tank Mk II, Matilda I—2pdr gun. 1940.

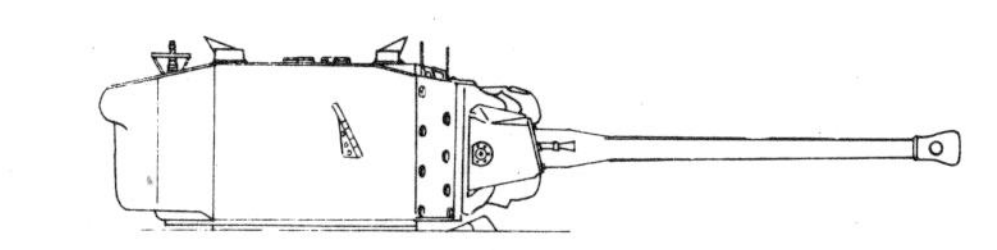

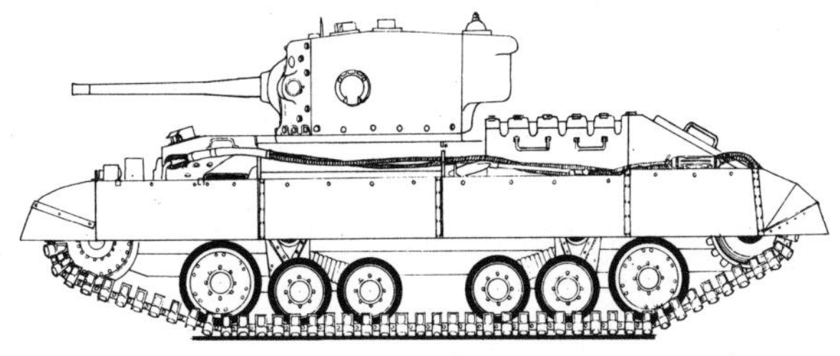

554. Infantry Tank Mk III, Valentine II—2pdr gun. 1941.
(inset: turret for Valentine XI, 75mm gun).

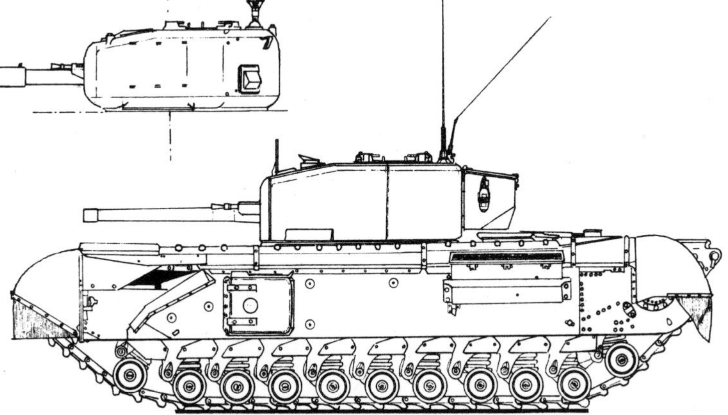

555. Infantry Tank Mk IV, Churchill III—6pdr gun. 1942.
(inset: turret for Churchill V, 95mm howitzer).

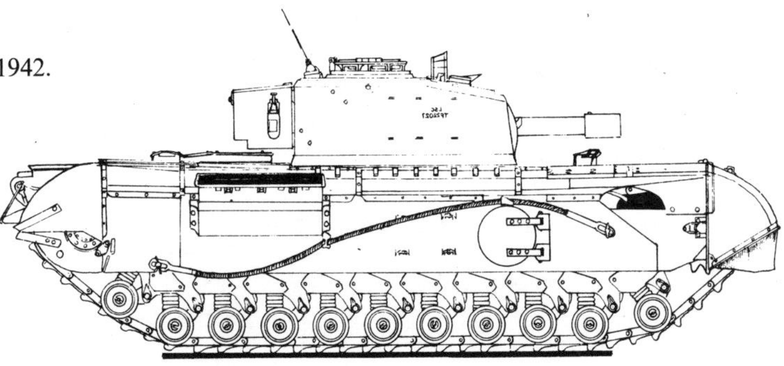

556. Infantry Tank, Churchill VIII—95mm howitzer. 1944.
(Churchill VII similar but with 75mm gun).

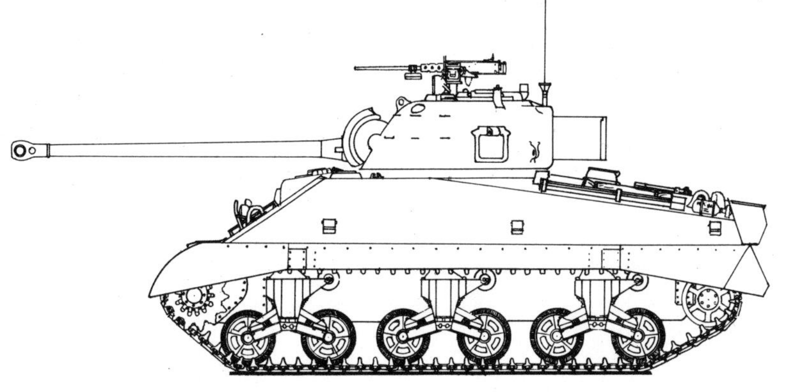

557. Sherman VC, Firefly—17pdr gun. 1944.

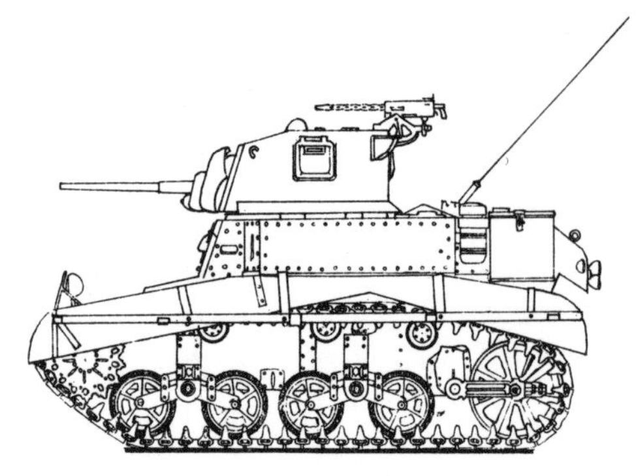

558. Light Tank M3A1 (Stuart III)—37mm gun, 1942. (also used by British).

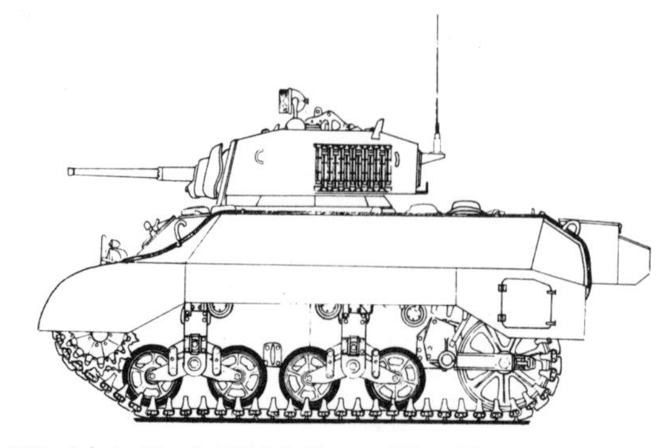

559. Light Tank M3A3 (Stuart V)—37mm gun. 1942. (also used by British).

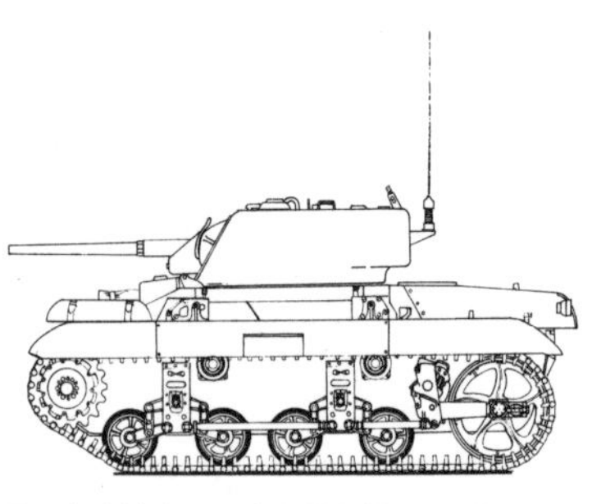

560. Light Tank (Airborne) M22 (Locust)—37mm gun. 1943. (also used by British).

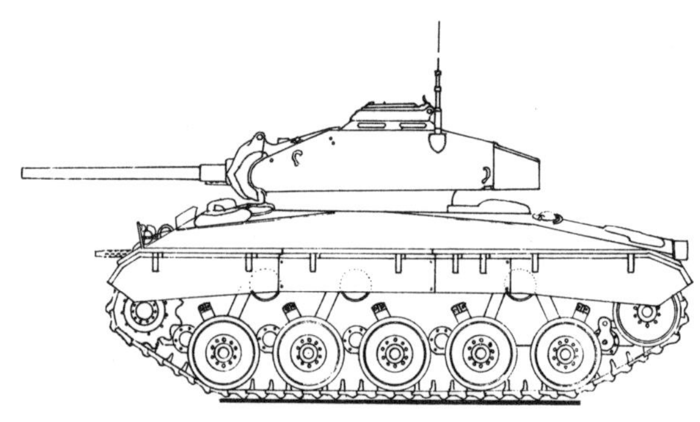

561. Light Tank M24 (Chaffee)—75mm gun. 1944. (also used by British).

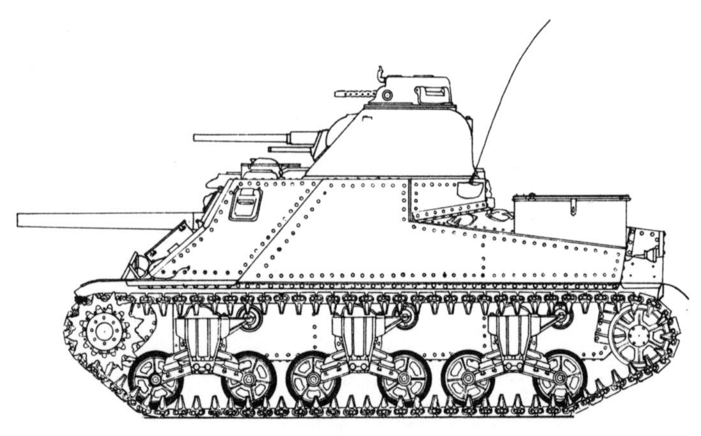

562. Medium Tank M3A5 (Grant II)—75mm gun. 1941. (also used by British).

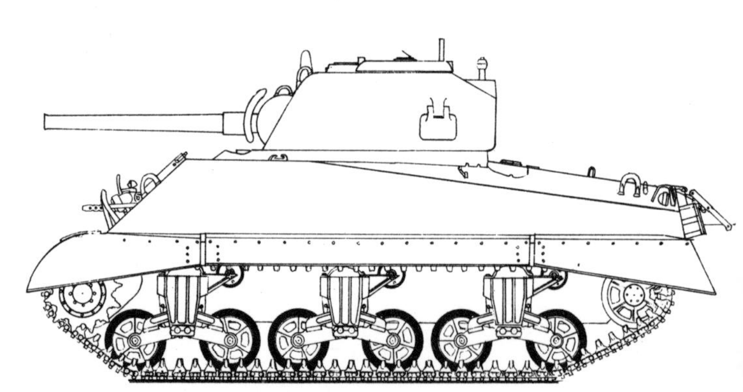

563. Medium Tank M4A3 (Sherman IV)—75mm gun. 1942. (also used by British).

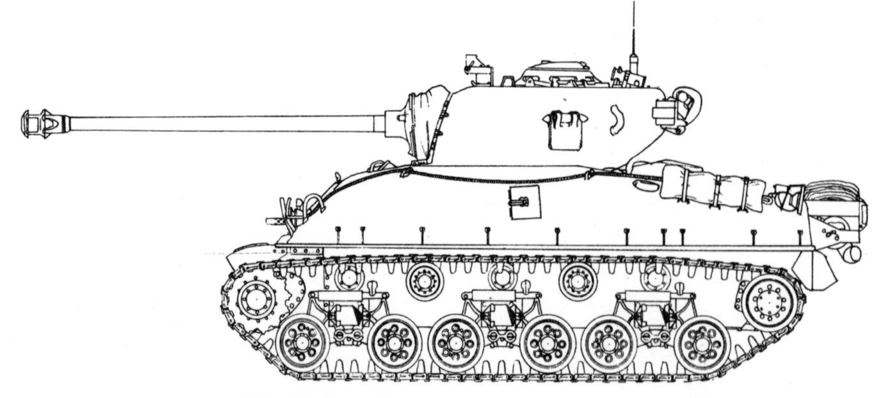

564. Medium Tank M4A1 (76mm) HVSS—76mm gun. 1944.

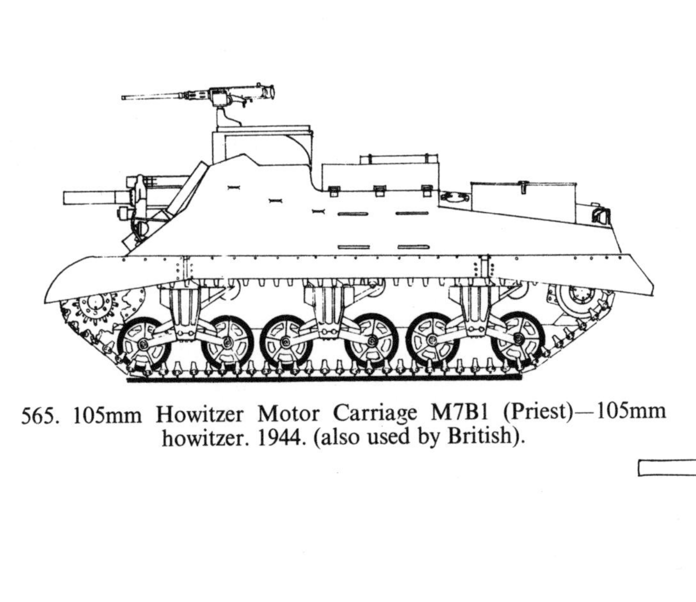

565. 105mm Howitzer Motor Carriage M7B1 (Priest)—105mm howitzer. 1944. (also used by British).

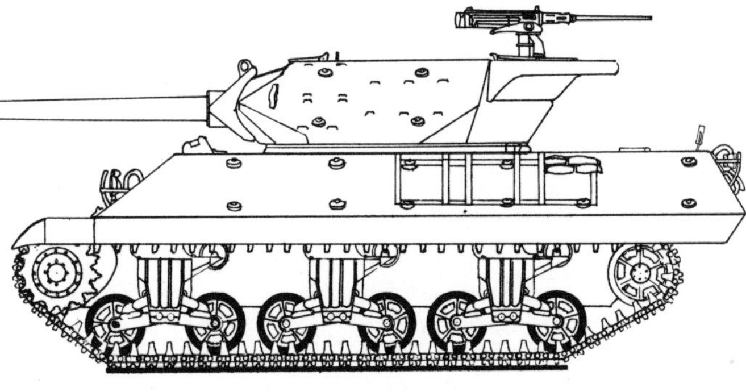

566. 3in Gun Motor Carriage M10 (Wolverine)—3in gun. 1944 (also used by British, and with 17pdr gun as Achilles).

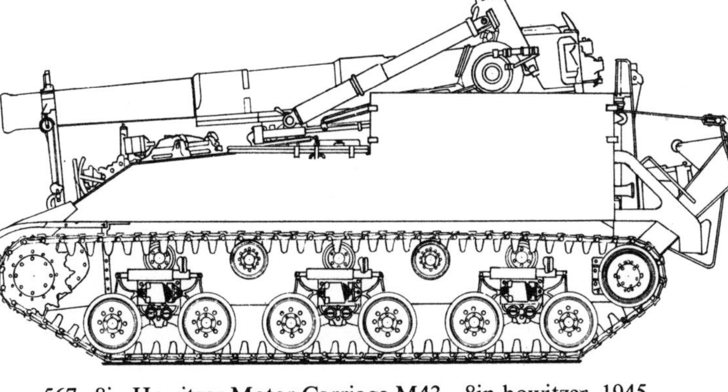

567. 8in Howitzer Motor Carriage M43—8in howitzer. 1945. (M40 similar but with 155mm gun).

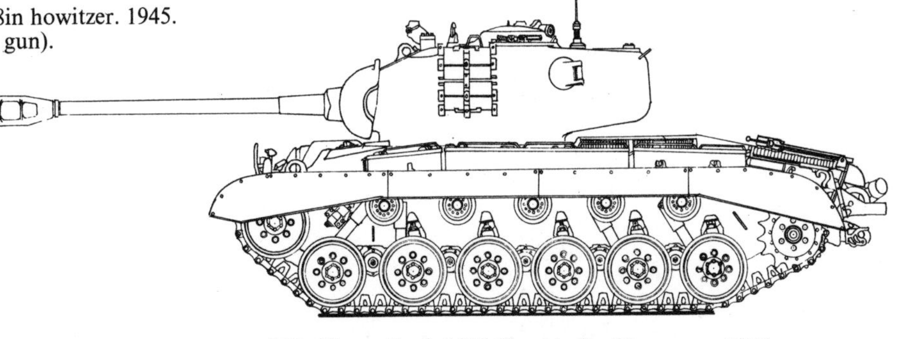

568. Heavy Tank M26 (Pershing)—90mm gun. 1945.

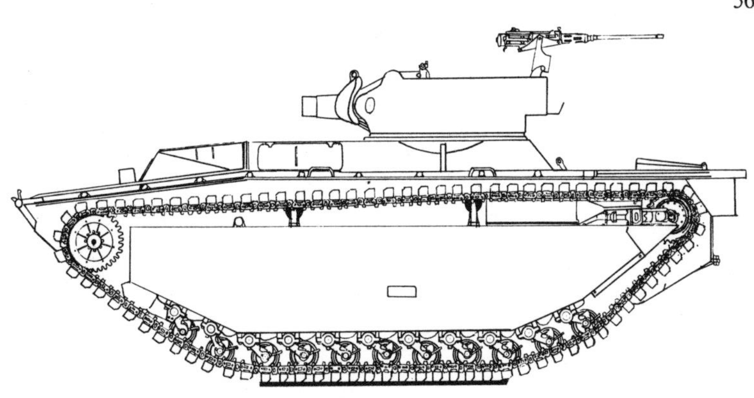

569. Landing Vehicle Tracked (Armoured) 4—75mm howitzer. 1944. (LVT(A) 4).

APPENDIX 3

Ordnance Designations of British Tanks ('A' numbers).

A1 (A1E1)	"Independent"; pilot model only.
A2E1	Medium Tank Vickers Mk I 1924.
A2E2	As above, CS version.
A3E1	3-Man Tank (ROF) 1925.
A4	Vickers "Carden-Loyd" light tank series, 1929 onwards. A4E13-18 (Mk II variants), A4E19-20 (Mk IV variants)
A5	Vickers "Carden-Loyd" 3-man light tank. Pilot model only.
A6E1, E2, E3.	Experimental 16 ton tanks (Medium Mk III) 1928.
A7E1, E2, E3.	Experimental medium tanks designed and built in ROF Woolwich.
A8 (A8E1)	Experimental medium tank ordered from Vickers. 1934-37.
A9 (A9E1)	Cruiser Tank Mk I, 1935.
A10 (A10E1, E2, E3)	Cruiser Tank Mk II. 'Valentine' developed from A10E3.
A11 (A11E1)	Matilda Infantry Tank Mk I, 1936-38.
A12 (A12E1)	Matilda Infantry Tank Mk II, 1938-39.
A13 Mk I (A13E1, E2, E3)	Cruiser Tank Mk III, 1937. (A13E1 was Christie vehicle purchased from USA).
A13 Mk II	Cruiser Tank Mk IV, 1937.
A13 Mk III	Cruiser Tank Mk V "Covenanter" 1937.
A14 (A14E1, E2)	Modified GS Specification Heavy Cruiser Pilot abandoned after trials
A15 (A15E1)	Crusader, Cruiser Tank Mk VI, 1938-40.
A16 (A16E1)	Development of A13 as Heavy Cruiser. Pilot model only 1938.
A17 (A17E1)	Light Tank Mk VII Tetrarch, 1939.
A18	Projected Cruiser Tank based on A17, 1939.
A19	Projected Cruiser Tank with auxiliary turrets on top of main turret. Abandoned 1939.
A20 (A20E1, E2)	Shelled Area Infantry Tank. Two pilots built. Cancelled in favour of A22.
A21	Projected development of A20. Drawing board project only.
A22 (A22E1)	Infantry Tank Mk IV Churchill.
A22F	Churchill VII (major design modifications).
A23	Projected lighter version of A22. Drawing board project.
A24 (A24E1)	Pilot of Cruiser Tank Mk VII Cavalier 1941.
A25 (A25E1)	Harry Hopkins, Light Tank Mk VIII 1941.
A26	Projected lighter and faster version of A22.
A27	A 27 L Centaur A 27 M Cromwell; Cruiser Mk VIII Tank.
A28	Cromwell (A27M) with increased armour and skirting plates over suspension.
A29	Large cruiser tank to carry 17pdr gun. Abandoned at project stage. Rolls-Royce design.
A30	(1) Challenger. (2) Avenger SP
A31	Cromwell with heavier armour. Project only.
A32	Cromwell with armour increased to standard of A22. Rolls-Royce design. Project only.
A33	Pilot assault tank. Built by English Electric 1942-43.
A34	Comet 1942-44.
A35	Heavier version of Cromwell. Project only.
A36	A30 with increased protection, stronger suspension and 17pdr gun. Project only.
A37	Heavier version of A33 with an extra suspension bogie per side, longer hull and 17pdr gun. Project only.
A38	Infantry Tank Valiant I and Valiant II
A39	Tortoise Heavy Assault Tank.
A40	Heavier version of A30. Project only.
A41 (A41A)	Centurion I, 1945.
A42	Churchill VII (A22F), redesignated, 1945.
A43	Infantry Tank, Black Prince, or "Super Churchill"
A44	Project for "Comet" with larger turret ring to take 17pdr gun. Cancelled.
A45	Centurion series Universal chassis for projected adaptation to several roles. Completed 1946 and with modifications became basis of Caernarvon/Conqueror.

Note: Designations in parentheses indicate prototype or pre-production vehicles, not necessarily actually built.

APPENDIX 4

Select Bibliography

Most of the reference sources for this book were drawn from unpublished documents, reports, cuttings, memoranda, vehicle handbooks and other archive material mainly in the Imperial War Museum and Royal Armoured Corps Museum libraries, plus United States Ordnance Department notes. The following published works include important coverage of various aspects relevant to the subject.

Bellona Military Vehicle Prints (Bellona Publications, Bracknell, Berks.) 1963–69.

BINGHAM, MAJOR JAMES. *Cromwell Mk IV* (Profile Publications, Surrey) 1968.
Infantry Tank Mk II (Profile Publications, Surrey) 1968.

CHAMBERLAIN, P. AND ELLIS, C. *The Sherman* (Arms and Armour Press, London and Arco, New York) 1968.
M4A3 Sherman (Profile Publications, Surrey) 1968.
Light Tank Mk VII Tetrarch (Profile Publications, Surrey) 1968.
Light Tanks M1–M5 Series (Profile Publications, Windsor, Berks.) 1969.

DUNCAN, N. W. *Light Tanks Mk I–VI* (Profile Publications, Windsor, Berks.) 1969.

GREEN, C. M., THOMSON, H. C. AND ROOTS, P. C. *Planning Munitions for War: Ordnance Department. United States Army in World War II: The Technical Services* (Department of the Army, Washington, D.C.) 1955.

HALL, H. DUNCAN AND WRIGLEY, C. C. *Studies of Overseas Supply. History of the Second World War* (H.M.S.O., London) 1955.

ICKS, LIEUT.-COL. ROBERTS J. *Tanks and Armored Vehicles* (Duell, Sloane and Pearce, New York) 1945.
M24 Chaffee (Profile Publications, Surrey) 1968.
M47 Patton (Profile Publications, Surrey) 1968.

LIDDELL HART, CAPTAIN B. H. *The Tanks, A History of the Royal Tank Regiment, 1914–45.* 2 volumes (Cassell, London) 1959.

NORMAN, CAPTAIN MICHAEL. *Centurion 5* (Profile Publications, Surrey) 1968.

OGORKIEWICZ, R. M. *Armour* (Stevens, London and Praeger, New York) 1960.
Design and Development of Fighting Vehicles (Macdonald, London and Doubleday, New York) 1968.

POSTAN, M. M. *British War Production. History of the Second World War* (H.M.S.O., London) 1952.

——HAY, D. AND SCOTT, J. D. *Studies in Government and Industrial Organisation: Design and Development of Weapons.* (H.M.S.O., London) 1964.

ROYAL ARMOURED CORPS TANK MUSEUM. *Movement and Firepower* (R.A.C. Publications, Bovington, Dorset) 1960.
Tanks 1919–39, The Inter-war period (R.A.C. Publications, Bovington, Dorset) 1960.
Tanks 1940–46, The Second World War (R.A.C. Publications, Bovington, Dorset) 1960.

SCOTT, J. D. AND HUGHES, R. *The Administration of War Production.* (H.M.S.O., London) 1955.

SELECT COMMITTEE, Report by. *Wartime Tank Production* (H.M.S.O., London) 1945.

STETTINIUS, E. R., Jr. *Lease-Lend* (Penguin, London) 1944.

TIMES, THE. *British War Production, 1939–45.* (The Times, London) 1945.

THOMSON, H. C. AND MAYO, L. *Procurement and Supply: Ordnance Department, United States Army in World War II: The Technical Services* (Department of the Army, Washington, D.C.) 1960.

VANDERVEEN, B. H. *M3 Half-track APC* (Profile Publications, Surrey) 1968.

WHITE, B. T. *British Tanks, 1915–45* (Ian Allan, Shepperton, Middx.) 1963
Churchill, British Infantry Tank Mk IV (Profile Publications, Windsor, Berks.) 1969.
Valentine, British Infantry Tank Mk III (Profile Publications, Windsor, Berks.) 1969.

APPENDIX 5

Index

As stated in the Preface, this index is arranged by function, all vehicles being grouped in classes (light tanks, cruiser tanks, armoured recovery vehicles, bridgelayers, etc.), so that the reader wishing to find reference to vehicles of a particular type can find here a listing of everything within that category together with page numbers of text entry and the number of the plate. The plate numbers are set in **bold** type after the page numbers. Some vehicles are listed in this index more than once since a few major production types developed for special functions are included with the basic chassis as well as within the special function grouping. This is mainly for the convenience of readers who may, for example, look for the Churchill AVRE under the "infantry tank" grouping instead of under the less obvious "assault engineer vehicle" grouping. It should be noted that some development vehicles mentioned in the text, mainly those built prior to 1939, are not illustrated in this book since they played no part in combat or AFV development in World War II itself. These vehicles are, however, included in the index. Where no illustration is given, this is indicated by a dash (—). American AFVs also used by the British (or any other Commonwealth nation) are indicated by (B) immediately after the vehicle designation with the British designation (if different) following the American designation. British mark numbers, if allocated retrospectively, are shown in brackets. The notation "(test)" indicates a "one off" trials vehicle. Similarly the notation "(project)" indicates a proposed vehicle which did not proceed beyond the design (or sometimes mock-up) stage.

The sequence of the vehicles in the index, under their function classification, does not follow alphabetical order but the order of the development of the vehicles as far as is practical. Thus this index is also a check-list of vehicles in chronological order.

Wherever possible, illustrations in this book are located adjacent to the description of the relevant vehicle in the text, but in some cases, where a large number of variants are involved, the picture may appear one or more pages away from the text entry. As a general rule, however, the illustration appears on or near the page number given.

British Light Tanks

American Light Tanks (and Combat Cars)

New Zealand Light Tank

British Cruiser Tanks (and Medium Tanks)

American Medium Tanks

Australian Cruiser Tanks

Canadian Cruiser Tanks

British Infantry Tanks (and Heavy Tanks)

American Heavy Tanks (and Assault Tanks)

Miscellaneous Tank Types

British Self-Propelled Guns (on tank chassis) (SP Guns)

(for AA tanks, *see* separate heading)

American Self-propelled Guns (Motor Carriages) (on Tank Chassis)

(GMC: gun motor carriage,
HMC: howitzer motor carriage,
MMC: mortar motor carriage)

Canadian Self-propelled Guns on Tank chassis

British AA (Anti-Aircraft) Tanks

American AA (Anti-Aircraft) Tanks

Canadian AA (Anti-Aircraft) Tanks

Miscellaneous Special Purpose Types, Projects, and Test Vehicles

Landing Vehicles Tracked (LVT)

American Half-tracks and Half-track Motor Carriages

(GMC: gun motor carriage,
HMC: howitzer motor carriage,
MMC: mortar motor carriage)

British Infantry Carrier Chassis with Self-propelled Weapons

Principal Tank Guns

(British)

(American)

Principal Tank Engines

(British)

(American)

Principal Suspension Systems

(British)

(American)

Miscellaneous